THE

LAW RELATING TO DOGS

BY

FREDERICK LUPTON,

SOLICITOR OF THE SUPREME COURT.

THE LAWBOOK EXCHANGE, LTD.
Clark, New Jersey
2007

ISBN-13: 978-1-58477-771-7
ISBN-10: 1-58477-771-0

The quality of this reprint is equivalent to the quality of the original work.

THE LAWBOOK EXCHANGE, LTD.
33 Terminal Avenue
Clark, New Jersey 07066-1321

Please see our website for a selection of our other publications and fine facsimile reprints of classic works of legal history:
www.lawbookexchange.com

Library of Congress Cataloging-in-Publication Data

Lupton, Frederick.
The law relating to dogs / Frederick Lupton.
p. cm.
Originally published: London : Stevens and Sons, 1888.
Includes bibliographical references and index.
ISBN-13: 978-1-58477-771-7 (cloth : alk. paper)
ISBN-10: 1-58477-771-0 (cloth : alk. paper)
1. Dogs--Law and legislation--Great Britain. I. Title.
KD667.D6L87 2007
346.4104'7--dc22

2006035844

Printed in the United States of America on acid-free paper

THE

LAW RELATING TO DOGS

BY

FREDERICK LUPTON,

SOLICITOR OF THE SUPREME COURT.

LONDON:

STEVENS AND SONS, 119, CHANCERY LANE,

Law Publishers and Booksellers.

1888

PREFACE.

So far as I am aware, no treatise has yet appeared dealing specifically with the Law relating to Dogs, which, nevertheless, forms a subject of no slight importance, especially when regard is had to recent legislation and decisions. I have, therefore, been induced to prepare the present little Work, in the hope that it may prove useful as one of reference both to the profession and the general public. Whether I have succeeded in my endeavours I must leave for the determination of my readers, while craving their kind indulgence for all errors and imperfections which may have crept in.

The provisions of some of the Statutes relating to Dogs are considered somewhat fully in the Text, whilst others will be found set forth *verbatim* in the Appendix.

I have avoided the use of technical expressions as far as possible, in order to render the Work more acceptable to those who are not members of the legal profession, and trust it will be found as complete and exhaustive as the limited range of a by no means abstruse branch of the Law will permit.

I have to acknowledge my obligations to the Hon. Baron Pollock for kindly perusing the proof sheets and offering several valuable suggestions.

FRED[K] LUPTON.

GREAT YARMOUTH,
May, 1888.

CONTENTS.

CHAPTER VII.

CHAPTER VIII.

CHAPTER IX.

CHAPTER X.

APPENDIX.

Statutes.

TABLE OF CASES CITED.

THE

LAW OF DOGS.

CHAPTER I.

INTRODUCTION.

FOR the purposes of this treatise dogs may be conveniently classified under four distinct heads, namely:—(1) ferocious; (2) dangerous; (3) mischievous; and (4) harmless. There is seldom much practical difference in the first three, as they usually display the same traits of character, and the rules of law applicable to them are almost precisely similar in each instance. We shall, therefore, speak of them collectively, except where circumstances render a departure from this course necessary or advisable, as in the chapters on the "Statutes relating to Dangerous Dogs," and "Trespassing Dogs." Properly speaking, however, a *ferocious dog* is one of a naturally savage disposition, of which he has already given proof by attacking or biting men or animals. Such a dog is also, clearly, both dangerous and mischievous. A *dangerous dog* is generally of a like nature, but not necessarily so, though the situation in which he is placed, or his condition at certain times, may render him extremely dangerous. Such would be a dog guarding a house, or carrier's cart; or which is suffering from rabies, or has been bitten by another dog suffering from that disease. A *mischievous dog* is commonly classed with such as are ferocious and dangerous, but is, in strictness, one whose destructive or predatory instinct prompts him to commit damage by jumping on people; or to injure property by spoiling crops, or by pursuing fowls, game, &c. A *harmless dog* is one of a tame and gentle nature, not given to commit acts either of ferocity or mischief; and in this light, as we shall see presently, every dog is regarded in the eye of the law until the contrary has been proved.

Distinction between ferocious

Before we proceed to consider the main portion of our subject, and in order that we may rightly comprehend the same, we must be careful to notice an important distinction which exists between animals naturally ferocious and dangerous, such as lions, bears, &c.,

and those of a domestic nature, which are ordinarily tame, docile, and inoffensive, such as horses, oxen, *dogs*, &c. It is laid down, that whoever keeps an animal accustomed to attack or bite mankind is, in general, liable to an action for damages at the suit of any person attacked or injured by it, the gist of the action being the keeping of the animal after knowledge of its ferocious disposition (*a*). **and domestic animals.**

In an action for damages against the owner of an animal naturally ferocious, it is not necessary for the person who has been attacked or injured to prove any default in the securing or taking care of the animal, since its dangerous and ferocious disposition must be well known to everybody; neither is it any defence to show that it was tame up to the time of the attack being made, though that circumstance may, perhaps, be pleaded in mitigation of damages. A person who keeps such an animal is bound to keep it so that it shall do no damage, and if it be insufficiently kept, or so kept that a person

(*a*) *May* v. *Burdett*, 9 Q. B. 101; *Jackson* v. *Smithson*, 15 M. & W. 565; 15 L. J. Ex. 311; *Besozzi* v. *Harris*, 1 F. & F. 92.

passing is not sufficiently protected, the owner is liable (*b*). But dogs and other domestic animals are presumed to be, as already stated, of a tame and gentle disposition. When, therefore, such an animal breaks through the ordinary tameness of its nature, and inflicts injury on person or property, it is, subject to a few exceptions, which we shall notice in due course, necessary for the aggrieved party, in order that he may successfully maintain an action for damages, to prove what is technically termed a "scienter," or, in other words, that the owner of the animal knew and was well aware of its dangerous or mischievous propensity (*c*). Though this rule is obviously consonant neither with moral justice or common sense; while the unfairness and absurdity of it has been severely commented on, from time to time, by learned judges and others; while a different and more reasonable doctrine prevails in other countries, where the mere fact of keeping the animal is sufficient to impose responsibility on the owner if it commits any damage; the law of England requires, to this day,

Scienter.

(*b*) *Besozzi* v. *Harris*, *supra*.
(*c*) See further on this subject, in Chapter II., p. 7.

evidence of scienter in nearly all instances of injuries by dogs. A number of cases were formerly decided in which the owners of sheep and cattle were held not entitled to recover damages for injuries so inflicted; but in the year 1865 a step was taken in the right direction, by the passing of the 28 & 29 Vict. c. 60, whereby the owners of dogs have been rendered liable for injuries occasioned by such animals to sheep and cattle (including horses) without the necessity of proving a scienter (*d*). It is to be hoped that the time is not far distant when the necessity for it will be finally abolished in all cases; but, in the meantime, and until this very desirable amendment of the law takes place, sheep and cattle will continue to occupy a more favoured position than human beings in this respect. We must, however, view the state of the law as it exists at the present time on this subject, and in the next chapter shall, therefore, consider more in detail the nature and effect of the scienter; when evidence of it will be required, and the sufficiency or insufficiency of such evidence.

No scienter now necessary in cases of injury to sheep or cattle. (28 & 29 V. c. 60.)

(*d*) See *post*, p. 12.

CHAPTER II.

ACTIONS FOR INJURIES BY DOGS—EVIDENCE FOR PLAINTIFF AND DEFENDANT—SCIENTER—DAMAGES—MISCHIEVOUS DOGS.

In order to successfully maintain an action for damages sustained through the bite of or other injuries caused by a dog, the plaintiff must show—

(1) *That the defendant kept or harboured the dog.*

The mere permitting a dog to be on or resort to premises is a sufficient keeping of him to support the action (*e*). And by the 28 & 29 Vict. c. 60 (Injuries to Sheep and Cattle), s. 2, the occupier of the premises where the dog was kept or permitted to be at the time when the injury was committed, is to be deemed the owner of the dog and liable as such, unless he can prove that he

Ownership of dog.

(*e*) *M'Kone* v. *Wood*, 5 C. & P. 2; *Smith* v. *Great Eastern Rail. Co.*, L. R. 2 C. P. 4.

was not the owner at the time of the injury, and that the dog was kept or permitted to be on the premises without his sanction or knowledge; and that where there are more occupiers than one, the occupier of that particular part of the premises in which the dog was kept or permitted to remain at the time of the injury, is to be deemed the owner of the dog. Therefore, as soon as a dog which is in the habit of frequenting premises is known to be dangerous, the occupier should send him away or cause him to be destroyed (*f*).

The defence may consist of a denial on the part of the defendant that he kept or harboured the dog, and if he can show that he has done everything reasonable to expel the animal from his premises, he will not be liable (*f*).

(2) *It must be shown that the dog was of a fierce and mischievous nature, to the knowledge of the defendant.*

It is not sufficient to show that the dog was of a savage and mischievous nature, but Scienter.

(*f*) *Idem.*

it must also be shown that the defendant knew and was well aware of such nature in the animal (*h*). This is, as we have already seen in Chapter I., page 4, what is technically termed, proving a scienter.

In order to prove a scienter evidence may be given of former complaints having been made of the animal's ferocity or mischief to the defendant, whose conduct and any statements he may have made concerning the matter on such occasions should also be taken into account; but when it can be shown that the savage nature of a dog was known to the defendant, the plaintiff need not show that anybody before him had been bitten (*i*); for when once this knowledge on the part of the former has been clearly established, the same liability attaches to him as to the owner of a lion or a bear, and in either case it is his duty to watch and secure the animal, which he must thereafter keep at his peril (*j*).

(*h*) *Mason* v. *Keeling*, 12 Mod. 332; *Beck* v. *Dyson*, 4 Camp. 198; *Fleming* v. *Orr*, 25 L. T. 73; 1 W. R. 339; *Sanders* v. *Teape*, 51 L. T. R. (N. S.) 263; *Worth* v. *Gilling*, L. R. 2 C. P. 1.

(*i*) *Besozzi* v. *Harris*, 1 F. & F. 92; *ante*, p. 3.

(*j*) *Fleming* v. *Orr*, *supra*; *Charlwood* v. *Greig*, 3 C. & K. 48.

If it can be shown that the defendant has warned a person to beware of a dog lest he should be bitten, this will be evidence of the defendant's knowledge of the savage nature of the animal (*k*).

And a report of a dog having been bitten by a mad dog is evidence that the defendant knew the dog was dangerous, and ought to be confined; especially if, by tying him up, he has shown suspicion or knowledge of the danger (*l*).

And where it was shown that a dog had bitten the defendant's cook and a drayman, tearing his leather apron and jacket off his back, and that he complained to the defendant, who said it had "pinched" two or three persons in the course of the last six years, the jury found a verdict for the plaintiff (*m*).

Promise to make compensation for injury.

But a promise by the owner, on complaint being made to him of the injury done, to make compensation, is insufficient, though

(*k*) *Judge* v. *Cox*, 1 Stark. 285.

(*l*) *Jones* v. *Perry*, 2 Esp. 482.

(*m*) *Herrett* v. *Macdonald*, before Mr. Justice Grove, at Westminster, May 4th, 1880.

some evidence of scienter, as it may have been made purely from charitable and praiseworthy motives, and not as admitting legal liability in the matter (*n*). And in an action for knowingly keeping a fierce dog, which had bitten or wounded the plaintiff, proof of a mere habit of bounding upon and seizing persons, not so as to hurt them, though causing some annoyance and trivial accidental damage to clothes, would not sustain the action (*o*).

Dog bounding upon persons.

It is not necessary that the scienter should be brought home to the defendant personally. If knowledge of a dog's ferocity be communicated to a servant having charge of the animal, that will be sufficient (*p*); but merely washing and feeding the animal does not constitute having charge of it (*q*). If the communication be made to the wife or servant of the defendant, there is a presumption in law that the fact has been communicated

(*n*) *Beck* v. *Dyson*, 4 Camp. 198; *Thomas* v. *Morgan*, 2 C. M. & R. 496; 1 Gale, 172; 5 Tyr. 1085; *Sanders* v. *Teape*, 51 L. T. R. (N. S.) 263.

(*o*) *Line* v. *Taylor*, 3 F. & F. 731.

(*p*) *Baldwin* v. *Casella*, 7 L. R. Ex. 325.

(*q*) *Colgett* v. *Norris*, Q. B. Div., March 13th, 1886.

to him (*r*). So, where complaint was made to the wife of the defendant, who occasionally assisted him in his business of a milkman, it was held that there was evidence of scienter, inasmuch as the information was intended for the owner, and, as it was the duty of the wife to communicate it to her husband, it might be presumed that she had done so (*s*). But, although notice to the wife will be primâ facie sufficient to charge the husband, the converse does not hold, and a notice to the husband during his lifetime will not, taken alone, be sufficient proof to render the wife liable after his death. A man may be told many things which he never communicates to his wife (*t*). But where two persons, who had upon previous occasions (one of them twice) been attacked by a dog, were called to prove that they had gone to the defendant's public-house, and made complaint to two persons who were behind the bar serving customers,

Notice to husband during lifetime.

(*r*) *Gladman* v. *Johnson*, 36 L. J. C. P. 153; 15 L. T. (N. S.) 476; *Colgett* v. *Norris*, *supra*; compare with *Baldwin* v. *Casella*, *supra*.

(*s*) *Gladman* v. *Johnson*, *supra*.

(*t*) Per Mellor, J., in *Miller* v. *Kimbray*, 16 L. T. (N. S.) 360.

and that one of them had also complained to the barmaid, but there was no evidence that these complaints were communicated to the defendant, nor was it shown that either of the two men spoken to had the general management of the defendant's business or the care of the dog, it was held by Lord Coleridge, C. J., and Keating, J. (Brett, J., dissenting), that there was evidence of scienter (*u*). It must be observed that evidence of a communication only makes a primâ facie case, and the presumption in law may be rebutted (*v*).

As regards notice there is no difference between a corporation and an individual, and whatever is notice to a person competent to receive it is notice to the corporation (*w*).

A dog may be brought into court, and shown to the jury to assist them in judging of his temper and disposition (*x*).

As already stated in Chapter I., p. 5, the

(*u*) *Applebee* v. *Percy*, 9 L. R. C. P. 647.

(*v*) *Colgett* v. *Norris*, *supra*.

(*w*) *Stiles* v. *Cardiff Navigation Co.*, 10 Jur. (N. S.) 1199; 33 L. J. Rep. (N. S.) Q. B. 310; *Smith* v. *Great Eastern Rail. Co.*, L. R. 2 C. P. 4.

(*x*) *Line* v. *Taylor*, 3 F. & F. 731.

28 & 29 Vict. c. 60 (*y*), renders the owner of every dog liable in damages (*z*) for injury done to any sheep or cattle by his dog; and it is not necessary to show a previous mischievous propensity in the animal, or the owner's knowledge of such propensity, or that the injury was attributable to neglect on his part (*a*). The term "cattle" here includes "horses and mares" (*b*). The operation of this act extends only to England and Wales, but somewhat similar provisions have been enacted with regard to Ireland by the 25 & 26 Vict. c. 59, as amended by the 30 & 31 Vict. c. 116; and with regard to Scotland by the 26 & 27 Vict. c. 100.

Award of arbitrator.

And as the award of an arbitrator is generally final and conclusive, the court will refuse to set it aside, unless it can be shown that the arbitrator has been guilty of misconduct, or that the award has been obtained by undue means, or for some other very cogent reason. Therefore, where upon a reference of all matters in difference, without plead-

(*y*) See the Act in Appendix, p. 127.

(*z*) As to the recovery of these damages, see *post*, p. 19.

(*a*) The Act, sect. 1.

(*b*) *Wright* v. *Pearson*, 38 L. J. (N. S.) Q. B. 312; L. R. 4 Q. B. 582. See also *Cowell* v. *Mumford*, 3 Times Law Rep. 1.

ings, the arbitrator found that the defendant had been guilty of negligence, in that his two greyhounds coupled together rushed against the plaintiff on the high road, knocked him down and broke his leg, it was held, that this finding was good, although there was no evidence of a scienter (*c*).

No scienter need be shown under sect. 2 of the 34 & 35 Vict. c. 56 (Dogs Act, 1871) (*d*).

(3) *The plaintiff must prove that he has been bitten, or has sustained actual damage from the ferocity or mischievousness of the animal.*

In an action to recover damages for the bite of a dog, medical evidence on behalf of the plaintiff will be most important, if not absolutely indispensable. In answer, the defendant may state that the plaintiff was a trespasser on the premises at the time of the occurrence, and this will be a good defence, if it can be shown that the plaintiff was upon the defendant's land, where he had no right to be, at the time of the occurrence; or if the injury

Defence that plaintiff was a trespasser when he received the injury.

(*c*) *Jones* v. *Owen*, 24 L. J. (N. S.) C. P. 587.
(*d*) See *post*, p. 23.

arose from the plaintiff's own carelessness, with knowledge of the danger (*e*). But a man has no right to keep a ferocious dog in such a situation, in the way of access to his house, that a person coming there for a lawful purpose in the daytime may be injured by it. And so with respect to a footpath, though it be a private one, a man has no right to put a dog with such a length of chain, and so near the footpath, that he could bite a person going along it (*f*). If no suspicion is thrown upon the injured party by the owner in such a case, it may be assumed that he had good cause for being on the owner's premises, provided the dog is put in a place forming an entrance to the house of the owner, although there may be other entrances to the house of a more public description, by which the injured party might have proceeded. And in such a case the putting up of a notice to "beware of the dog," will not exempt the owner from liability to the person injured, if it appears that the latter could not read, or, in fact, did not read the notice. Neither

Notice to beware of dog.

(*e*) Per Tindal, C. J., in *Sarch* v. *Blackburn*, 4 C. & P. 400; 1 M. & W. 505.

(*f*) *Idem.*: *Curtis* v. *Mills*, 5 C. & P. 489.

will a verbal caution to the same effect, if the accident was not occasioned by the carelessness or want of caution of the person injured (*g*). But a man has a right to keep a ferocious dog for the protection of his premises, and to turn it loose during the night, provided the barking of the dog does not disturb the rest of the neighbours and create a nuisance (*h*).

Right to keep ferocious dog.

Defence that plaintiff provoked dog.

The defendant may also plead that the injury was sustained by the plaintiff in consequence of his having provoked the animal. And if it can be shown that the dog has bitten or done damage, only because he has been provoked, the person who occasioned the injury that has happened will be accountable for it, and if he be the person who has sustained the injury, he is alone to blame. And so also, if the dog which has done the damage has been exasperated and stirred up by another dog, the master of the latter will be accountable for the damage; and if a dog is kept properly secured, and another person lets him loose and urges him to mischief (*i*),

(*g*) *Idem.*
(*h*) *Brock* v. *Copeland*, 1 Esp. 203; and see *post*, p. 76.
(*i*) This is an offence punishable summarily. See the

the owner is not liable. Therefore, it is not enough to show that a dog has killed sheep in order to entitle the owner to recover compensation from the master of the dog, because the dog may have been properly secured by the master, and improperly let loose and urged to mischief by a third person without his knowledge, and even against his express prohibition (*j*). If a person who is acquainted with the ferocious disposition of an animal lets it loose he will be responsible in damages for whatever injuries it may inflict (*k*).

Plaintiff need not allege negligence in keeping dog.

The plaintiff need not allege or prove any negligence or want of care in the keeping of the animal (*l*). In *Card* v. *Case* (*m*), Coltman, J., said, "The question is not, What is the wrongful act alleged? Certainly not the breach of duty in not carefully keeping the dog; that would be taking

10 & 11 Vict. c. 89, s. 28, and the 2 & 3 Vict. c. 47, s. 54, *post*, p. 51.

(*j*) *Fleming* v. *Orr*, 25 L. T. 73; 1 W. R. 339; 2 Macq. H. L. Cases, 14.

(*k*) *Leame* v. *Bray*, 3 East, 593.

(*l*) *May* v. *Burdett*, 9 Q. B. 101; *M'Kone* v. *Wood*, 5 C. & P. 2; *Jackson* v. *Smithson*, 15 M. & W. 565; 15 L. J. Ex. 311; *Smith* v. *Great Eastern Rail. Co.*, L. R. 2 C. P. 4.

(*m*) 12 Jur. 247; 17 L. J. C. P. 124.

a very narrow view of the words. I think if it were true that it was necessary to allege negligence on the part of the defendant, the scienter might possibly be inducement, but *May* v. *Burdett* (*n*) shows that the keeping or not keeping negligently is not an essential part of the complaint. The real grievance is the keeping with knowledge, and this, when followed by injury to the plaintiff, is the ground of action." And Maule, J., in the same case said, "The allegation here is that the defendant kept a ferocious dog, well knowing it to be so. The two cases referred to, *May* v. *Burdett*, and *Jackson* v. *Smithson* (*o*), as also the general course of precedents, show that the knowingly keeping a savage animal is in itself a wrongful act; and, therefore, any person injured may bring his action without alleging negligence" (*p*). If the allegations on the part of the plaintiff are proved, he will be entitled to recover damages for the bodily suffering which he

(*n*) *May* v. *Burdett*, 9 Q. B. 101, *supra*.

(*o*) *Jackson* v. *Smithson*, 15 M. & W. 565; 15 L. J. Ex. 311, *supra*.

(*p*) *Ante*, p. 3.

has endured, together with the expenses of medical or surgical attendance, and all such reasonable and necessary consequential expenses as he may have incurred.

The defendant may employ a surgeon on his own behalf to examine the nature of the injuries said to have been received by the plaintiff, and call upon him to give evidence either for the purpose of entirely disproving the plaintiff's case, or showing that the extent of the injuries sustained has been unduly exaggerated.

Proceedings must be instituted in the Queen's Bench Division of the High Court of Justice, unless the amount of the claim does not exceed 50*l.*, when the county court has jurisdiction. And by the 28 & 29 Vict. c. 60, s. 1 (*q*), damages are recoverable under the act in any court of competent jurisdiction by the owner of sheep or cattle (*r*) injured. And where the amount of the damage does not exceed *five pounds*, the same is recoverable in a summary way before any justice or justices sitting in petty sessions under the

(*q*) See the Act in the Appendix, p. 127.
(*r*) See *ante*, p. 13.

11 & 12 Vict. c. 43. These damages will be now recoverable as a civil debt under the 42 & 43 Vict. c. 49, ss. 7 and 35.

Mischievous dogs.

In actions for injuries occasioned to person or property by *mischievous* dogs, properly so called (*s*), the rules as to proving a scienter, the nature of the pleadings, and proceedings generally, will be very similar in all the most important respects to those where injuries have been caused by bites, but they will, of course, vary according to the circumstances of each particular case. The following instances will show the responsibility incurred by persons who keep *mischievous* dogs.

In *Sanders* v. *Teape*, the plaintiff, a labourer, was digging a hole in the garden of a house adjoining that of the defendant T. There was a small wall, only three feet high, between these gardens. This wall belonged to the defendant T. The plaintiff was engaged in doing some work at the bottom of the hole. Three dogs belonging to the defendant T. had been taken out by the other defendant S., and as the defendant S. was returning, the dogs ran through a gate into a garden

(*s*) See *ante*, p. 2.

adjoining the one where the plaintiff was at work. As the dogs were running about in playfulness, one of them, a large Newfoundland dog, jumped over the wall, and jumped or fell into the hole where the plaintiff was working at the time in a stooping posture. The dog fell on the nape of the plaintiff's neck, causing injuries through which he was confined to bed for three weeks, and he was unable to work for some time after. The defendant T. had offered the plaintiff a couple of sovereigns as compensation, which was refused. In an action for these injuries against the defendant T. as the owner of the dog, and against the defendant S. as having the dogs in charge, it was held that, inasmuch as the dogs were not shown to be mischievous to the knowledge of the owner, the plaintiff had no cause of action against either of the defendants(*t*).

If a dog *frightens* a horse and occasions damage by flying at, or running out and barking at it, the owner of the dog will be clearly liable if it can be shown that he was

(*t*) *Sanders* v. *Teape*, 51 L. T. R. (N. S.) 263, *supra*.

aware of the animal's propensity to such mischief.

The subject of dogs trespassing on land will be considered in Chapter IV.

CHAPTER III.

STATUTES RELATING TO DANGEROUS DOGS.

We must now notice the effect of the several statutes which have been enacted with the object of affording protection to the public against dangerous dogs (*u*), and also point out how and when the owners of such animals may be proceeded against under the statutes in courts of summary jurisdiction.

The operation of these statutes varies in extent. In some it extends to all parts of the kingdom; in others, to all parts of England, except the Metropolis; in others, again, to the Metropolis only. We will consider these in their order, and—

(1) *As to all parts of England.*

34 & 35 Vict. c. 56 (Dogs Act, 1871), s. 1.

By the 34 & 35 Vict. c. 56 (Dogs Act, 1871) (*v*), s. 1, the police may take possession of and detain dogs supposed to be savage or

(*u*) *Ante*, p. 14.

(*v*) This Act extends to all parts of the United Kingdom, and will be found set out verbatim in the Appendix, p. 129.

Stray dogs.

dangerous, and straying on the highway (*w*). Where the owner of any dog taken possession of is known, a letter, stating the fact of such dog having been taken possession of, is to be sent to the owner; and when any dog so taken has been detained for *three clear days* where he is not known, and for *five clear days* where he is known, without the owner claiming the dog, and paying all expenses incurred by its detention, the chief officer of the police of the district (*x*) in which the dog was found, may cause it to be sold or destroyed. All dogs detained must be properly fed and maintained at the expense of the local rate. And by sect. 2, "Any court of summary jurisdiction may take cognizance of a complaint that a dog is dangerous, and not kept under proper control; and, if it appears to the court that a dog is dangerous, the court may make an order in a summary way, directing the dog to be kept by the owner under proper control; and any person failing to comply with such order is liable to a

(*w*) Highway includes any street or place of public resort. Sect. 5.

(*x*) For meaning of the expression "police district," see the Act, sect. 5, *post*, p. 132.

penalty of *twenty shillings* for every day during which he fails to comply with such order." It is not necessary under this section to show knowledge on the part of the owner (*y*). The court may order a dangerous dog to be destroyed, without giving the owner the option of keeping it under control (*z*).

Dogs not under control.

And by sect. 3, "The local authority (*a*) may, if a mad dog, or a dog suspected of being mad, is found within its jurisdiction, make, and when made, vary or revoke, an order placing such restrictions as they think expedient on all dogs not being under the control of any person during such period as may be prescribed in such order throughout the whole of their jurisdiction, or such part thereof as may be prescribed in such order. Any person who acts in contravention of any order made in pursuance of this section is liable to a penalty not exceeding *twenty*

(*y*) Per Lord Coleridge, C. J., and Mathew, J., in *Parker* v. *Walsh*, Times, July 1, 1885.

(*z*) *Pickering* v. *Marsh*, 43 L. J. (N. S.) M. C. 143.

(*a*) Local authority here means the Corporation in boroughs and the Justices in Petty Sessions in counties. The Act, sect. 5.

shillings. Due notice of such order is to be published at the expense of the local rate (*b*). The provisions of the act as to the detention, and sale, or destruction of dogs found straying on the highway (*c*) apply to dogs found at large in contravention of any order made in pursuance of this section."

The question whether or not a dog is under "proper control" is one of fact, and not of law (*d*). And at length, after many conflicting decisions by magistrates, it has been settled that a dog is not under proper control, within the meaning of this section, unless muzzled or led. The control must be an effective, and not a moral, control; and, in the absence of any positive evidence to the contrary, the fact that a dog was neither muzzled nor led is sufficient to prove that it was not under proper control (*e*).

The Contagious Diseases (Animals) Acts,

(*b*) Evidence by a police constable that he had seen about six copies of a notice posted about a borough was held to be sufficient evidence that proper notice had been given. *Reg.* v. *Justices of Huntingdon*, L. R. 4 Q. B. Div. 522.

(*c*) Sect. 1, *ante*, p. 23.

(*d*) *Wren* v. *Pocock*, 34 L. T. (N. S.) 697; *Reg.* v. *Justices of Huntingdon*, *supra*; *Hay* v. *Bennett*, 3 Times Law Rep. 24.

(*e*) *Hay* v. *Bennett*, *supra.*

and the orders made by the Privy Council in pursuance thereof, have for their object the prevention and checking of such diseases amongst cattle and other animals. The principal act is that of 1878 (41 & 42 Vict. c. 74 (*f*)), which consolidates and repeals all former acts on the subject. Though it does not appear to have been originally intended to apply to dogs, yet since the passing of the Amending Act of 1886 (49 & 50 Vict. c. 32, s. 8 (*g*)), and the making of the Rabies Orders of the Privy Council, the principal act has now become of very great importance in its relation to them.

Contagious Diseases (Animals) Acts and Privy Council Orders.

By sect. 32 of the principal act, the Privy Council may make general or special orders for a variety of purposes; and by sect. 8 of the Amending Act of 1886 above mentioned, this power to make orders is declared to extend to the making of orders for extending the definition of animals in the principal act, so as to include any kind of four-footed beasts. Accordingly, under Article 7 of the

(*f*) This act extends to all parts of the United Kingdom.

(*g*) See *infra*.

Rabies Order, 1887, dogs are animals, and rabies a disease for the following purposes of the Acts of 1878 and 1886; namely:—Of the Act of 1878, sect. 31 (*Notice of Disease*); sect. 32 (*Orders of Council*); sect. 50 (*Powers of Police*); and sect. 51 (*Powers of Inspectors*). Of the Act of 1886, sect. 6 (*Slaughter*); and sect. 8 (*Definition of Animals*); and all other sections of those acts containing provisions relative to or consequent on the provisions of those sections, including such sections as provide for offences and procedure. We must now examine these sections, and also the articles of the Rabies Order, 1887.

By the combined operation of sect. 31 of the Act of 1878 and Article 8 of the Rabies Order, every person having in his possession or under his charge an animal affected with disease must, as far as practicable, keep it separate from animals not so affected, and with all practicable speed give notice of its being so affected to a police constable, who must himself forthwith give information to an inspector (*h*) of the local authority; and

(*h*) As to the appointment of inspectors, see the principal act, 41 & 42 Vict. c. 74, s. 42.

the inspector must in turn forthwith report the same to the local authority (*i*).

Optional notice of disease or illness.

Any person having in his possession or under his charge a dog, &c. affected with rabies or any illness, or suspected of being so affected, besides giving notice to a constable, as required by sect. 31 (*j*) of the Act of 1878, may give similar notice to the inspector of the local authority (*k*). An inspector of a local authority, on receiving information of the supposed existence of rabies, or having reasonable ground to sus-

Duty of inspector to act immediately.

(*i*) Local authorities in England are thus defined in Schedule 2 of the same Act.

1. Counties except within the Metropolis.	The justices in general or quarter sessions assembled.
2. The City of London and the liberties thereof.	The Corporation of London.
3. The Metropolis, except the City of London and the liberties thereof.	The Metropolitan Board of Works.
4. Boroughs subject to the Municipal Corporation Act, 1835.	The mayor, aldermen, and burgesses acting by the council.
5. Other boroughs.	The commissioners or other body maintaining the police therein.
6. The district of the local board of Oxford.	The local board.

See also sect. 9.

(*j*) *Supra.*

(*k*) Rabies Order, 1887, Article 9.

pect its existence, must proceed with all practicable speed to the place where such rabies exists, or is supposed to exist, and then and there put in force and discharge the powers and duties under the Acts of 1878 and 1886, and any Orders in Council conferred and imposed upon him as inspector (*l*).

Public warning as to existence of rabies.

The local authority may, from time to time, give public warning, by placards, advertisement, or otherwise, of the existence of rabies in any kennel, stable, building, field, or other place, with or without any particular description thereof, and continue to do so during the existence of the disease, and in case of a kennel, stable, building, or other like place, until the same has been cleansed and disinfected. It is unlawful without authority or excuse to remove or deface any such placard (*m*).

Special regulations of local authority as to dogs.

Any local authority may make regulations for the following purposes, or any of them:

(a) For providing for the muzzling of dogs, except when securely attached to or

(*l*) Rabies Order, 1887, Article 10.

(*m*) *Idem*, Article 11. As to penalty for defacing placard, see 41 & 42 Vict. c. 74, s. 61, *post*, p. 41.

kept within a kennel, stable, house, building, or other like place.

(b) For providing for the keeping of dogs under control (otherwise than by muzzling) by the owner or person in charge thereof, in such manner as may be prescribed by such regulations.

(c) For providing for the seizure and detention of stray dogs, or of dogs not muzzled, or of dogs not being kept under control, in the manner prescribed by such regulations, and for the recovery by the local authority of the expenses incurred by them in respect of the detention of any such dogs from the owners thereof.

(d) For providing for the seizure, detention, and disposal, including slaughter, of dogs being at large and affected with or suspected of rabies, or being or having been in the same kennel, stable, house, building, or other place, or in the same pack, or otherwise in contact, with dogs affected with rabies, or being or having been otherwise exposed to the infection thereof.

(e) For prohibiting or regulating the holding of shows or exhibitions of dogs, and the exposing of dogs for exhibition or sale thereat (*n*).

Regulations of local authority as to movement for rabies.

Also (a) for prohibiting or regulating the movement into or out of any kennel, stable, building, field, or other place of a dog, &c. affected with or suspected of rabies. (b) For prohibiting or regulating the movement into or out of any kennel, &c. in which rabies exists of a dog, &c. that has been in contact with, or in the same kennel, &c. with, a dog, &c. affected with rabies. (c) For prohibiting the taking out of any kennel, &c. of any fodder (*o*), litter (*p*), or other thing that has been in contact with or used for or about a dog, &c. affected with or suspected of rabies; but nothing in any such regulation is to authorize movement in contravention of any provision of any General Order of Council for the time being in force; and a regulation under paragraph B is to operate so long only as rabies exists in the judgment of the local authority in any kennel, &c. to

(*n*) Rabies Order, 1887, Article 12.

(*o*) "Fodder" includes meat, biscuits, or other substance commonly used for the food of dogs. *Idem*, Article 4.

(*p*) "Litter" means straw or other substance commonly used for bedding, or otherwise for or about dogs. *Idem*.

which the regulation refers, and in the case of a kennel, stable, building, or other like place, until the same has been cleansed and disinfected (*q*).

Regulations of local authority as to cleansing and disinfection.

Also (a) for providing for the cleansing and disinfection of places used by, and of utensils, feeding-troughs, pens, hurdles, or other things used for or about dogs, &c. affected with or suspected of rabies. (b) For providing for the cleansing and disinfection of vans, or carts, or other vehicles used for carrying dogs, &c. affected with or suspected of rabies on land, otherwise than by railway. (c) For prescribing the mode in which such cleansing and disinfection are to be effected. (d) For providing that such places, utensils, feeding-troughs, pens, hurdles or other things, vans, carts, and other vehicles should be cleansed and disinfected at the expense of the local authority, or of the owner, lessee, or occupier thereof. If any person fails to cleanse and disinfect any place, or any utensil, &c. in accordance with any such regulation, the local authority may, without prejudice to the recovery of any penalty for the infringement of such regulation, cause such

(*q*) Rabies Order, 1887, Article 13.

place, utensil, &c. to be cleansed and disinfected, and recover the expenses of such cleansing and disinfection from such person in any court of competent jurisdiction (*r*).

Obligation on owners and occupiers.

Where the power of causing any place or any utensil, &c. to be cleansed and disinfected is exercised by a local authority, the owner or occupier thereof must give reasonable facilities for that purpose (*s*).

Slaughter in rabies.

Where a person having a dog, &c. in his possession, or under his charge, gives notice to a constable that the dog, &c. is affected with rabies, or a person is convicted of an offence against the Act of 1878, by reason of his having failed to give such a notice (*t*) in respect of any dog, &c., then, if at any time thereafter it appears to the local authority, on a special report of a veterinary inspector (*u*), or veterinary surgeon, that the dog, &c. is affected with rabies, the local

(*r*) Rabies Order, 1887, Article 14. As to penalty for not cleansing, &c., see 41 & 42 Vict. c. 74, ss. 60 and 61, *post*, p. 40.

(*s*) *Idem*, Article 15. As to penalty for not doing so, see 41 & 42 Vict. c. 74, ss. 60 and 61, *post*, p. 40.

(*t*) *Idem*.

(*u*) Veterinary inspector means an inspector being a member of the Royal College of Veterinary Surgeons, or any veterinary practitioner qualified as approved by the Privy Council. *Idem*, s. 5.

authority may serve on the owner of the dog, &c. a notice, in writing, requiring him to slaughter it, or permit them to slaughter it forthwith (*v*). If, in any case, the owner fails to comply with the requisition of the notice of the local authority, he will be deemed guilty of an offence against the Act of 1878 (*w*), unless he shows to the court of summary jurisdiction before which he is charged that the dog, &c. was not affected with rabies (*x*).

Where a dog, &c. has died of or been slaughtered on account of rabies, or disease supposed to be rabies, the local authority may, previous to the disposal of the carcase, cause a post-mortem examination to be made thereof, in which case such examination is to be conducted by a veterinary inspector or veterinary surgeon specially appointed in that behalf, who must forthwith report to the local authority for their information the result of such examination. Where the power of causing a post-mortem examination is

(*v*) This is in pursuance of sect. 6 of the 49 & 50 Vict. c. 32 (repealing sect. 29 of the 41 & 42 Vict. c. 74).

(*w*) As to penalty, see 41 & 42 Vict. c. 74, ss. 60 and 61, *post*, p. 40.

(*x*) Rabies Order, 1887, Article 16.

exercised by the local authority, the owner or person in charge of the carcase must give all reasonable facilities for the purpose (*x*).

Disposal of carcases.

The local authority must cause the carcase (*y*) of every dog, &c. that has died of rabies, or been slaughtered in consequence of being affected with the disease, to be buried as soon as possible in its skin in some proper place, and to be covered with a sufficient quantity of quicklime or other disinfectant, with not less than six feet of earth: or the local authority may, if authorized by licence from the Privy Council, cause the carcase to be destroyed, under the inspection of the local authority, in the mode following. The carcase is to be disinfected, and then taken, in charge of an officer of the local authority, to a horse-slaughterer's or knacker's yard, approved for the purpose by the Privy Council, or other place so approved, and there destroyed by exposure to a high temperature, or by chemical agents. With a

(*x*) Rabies Order, 1887, Article 17. As to penalty for refusing, see 41 & 42 Vict. c. 74, ss. 60 and 61, *post*, p. 40.

(*y*) Carcase also means part of the carcase, and the flesh, bones, hide, skin, or other part of the animal, separately or otherwise, or any portion thereof. *Idem*, Article 4.

view to the execution of these provisions the local authority may make, from time to time, such regulations as they think fit for prohibiting or regulating the removal of any carcase, or for securing the burial or destruction of the same. Where a local authority causes a carcase to be buried, they must first cause its skin to be so slashed as to be useless. They may cause or allow a carcase to be taken into the district of another local authority to be buried or destroyed, with the previous consent of that local authority, or with a licence in that behalf of the Privy Council, but not otherwise (*z*).

Digging up carcases buried on account of rabies.

The digging up carcases of dogs buried on account of rabies is unlawful, except with the licence of the Privy Council (*a*).

Weekly returns to rabies.

Where an inspector of a local authority finds in his district rabies, he must forthwith make a return thereof to the local authority and to the Privy Council on a form provided by the Privy Council, with all particulars required, and continue to make a return

(*z*) *Idem*, Article 18.

(*a*) *Idem*, Article 19. As to penalty, see 41 & 42 Vict. c. 74, ss. 60 and 61, *post*, p. 40.

thereof on the Saturday of every week until the disease has ceased (*b*).

Production of licences, names, and addresses.

Every person in charge of dogs, &c. being moved, where under any regulation of a local authority a movement licence is necessary, must, on demand of a justice, constable, inspector, or other officer of the local authority, produce and show to him the movement licence, if any, authorizing the movement, and allow it to be read and a copy of or extract from it to be taken by the person to whom it is produced. Every person so in charge must on demand, as aforesaid, give his name and address to the justice, constable, inspector, or other officer (*c*).

By sect. 50 of the Act of 1878, the police of each police district, &c. must execute and enforce the act and every Order of Council.

Duties and authorities of constables.

Where a person is seen or found committing, or is reasonably suspected of being engaged in committing, an offence against the act, a constable may, without warrant, stop and detain him, and if his name and

(*b*) Rabies Order, 1887, Article 20.

(*c*) *Idem*, Article 22. See also Article 23, sub-sect. 6, *post*, p. 46, and (as to penalty for refusal) 41 & 42 Vict. c. 74, ss. 60 and 62, *post*, p. 40.

address are not known to the constable, and he fails to give them to the satisfaction of the constable, the constable may, without warrant, apprehend him; and the constable may, without warrant, whether so stopping, or detaining, or apprehending the person or not, stop, detain, and examine an animal, vehicle, boat, or thing, to which the offence or suspected offence relates, and require the same to be forthwith taken back to or into any place or district wherefrom or whereout it was unlawfully removed, and execute and enforce that requisition. If any person obstructs or impedes a constable or other officer in the execution of the act, or of an Order of Council, or of a regulation of a local authority, or assists in any such obstructing or impeding, the constable or officer may, without warrant, apprehend the offender. A person apprehended under the section must be taken with all practicable speed before a justice, and not detained without a warrant longer than is necessary for that purpose; and all enactments relating to the release of persons on recognizances taken by an officer of police or a constable are to apply in the case of a person apprehended under this section. The foregoing provisions respecting a constable extend and apply to any person called by a constable to his assistance. A constable must forthwith make a report in writing to his superior officer of every case in which he stops any person, animal, vehicle, boat, or thing under the section, and of his proceedings consequent thereon. Nothing in the section is to take away or abridge any power or authority that a constable would have had if the section had not been enacted.

Fines for offences.

By sect. 60, any person guilty of an offence against the act is liable to penalties, as follows, namely:—To a penalty not exceeding *twenty pounds.* If the offence is committed, with respect to more than *four* animals, to a penalty not exceeding *five pounds* for each animal; or where the offence is committed in relation to carcases, fodder, litter, dung, or other thing (exclusive of animals), to a penalty not exceeding *ten pounds* in respect of every half-ton in weight thereof after one-half ton, in addition to the first penalty of not exceeding twenty pounds.

General offences

By sect. 61, if any person, without lawful authority or excuse, does any of the following things, he will be guilty of an offence against the act, namely:—If he does anything in contravention of the act, or of an Order of Council, or of a regulation of a local authority. If where required by the act to keep an animal separate as far as practicable, or to give notice of disease with all practicable speed, he fails to do so. If he fails to give, produce, observe, or do any notice, licence, rule, or thing which, by the act or an Order of Council, or by a regulation of a local authority, he is required to give, produce, observe, or do. If he does anything which by the act or an Order of Council is made or declared not lawful. If he does or omits anything, the doing or omission whereof is declared by the act, or by an Order of Council, to be an offence by him against the act. If he refuses, to an inspector or other officer acting in execution of the act, or of an Order of Council, or of a regulation of a local authority, admission to any land, building, place, vessel, pen, vehicle, or boat, which the inspector or officer is entitled to enter or examine, or obstructs or impedes him in so entering or examining, or otherwise in any respect obstructs or impedes an inspector, or constable, or other officer, in the execution of his duty, or assists in any such obstructing or impeding. If he throws or places, or causes or suffers to be thrown or placed, into or in any river, stream, canal, navigation, or other water, or into or in the sea within three miles of the shore, the carcase of an animal which has died of disease, or been slaughtered as diseased, or suspected. And on a further conviction within a period of twelve months for a second or subsequent offence, he will be liable to be imprisoned for any term not exceeding one month, with or without hard labour, in lieu of the pecuniary penalty to which he is liable under the act (*d*).

Imprisonment instead of fine for use of expired licences, digging up of carcases, and other specified offences.

And by sect. 62: "If any person does any of the following things he will be guilty of an offence against the act: (1) If, with intent to unlawfully evade the act or an Order in Council, or a regulation of a local authority, he does anything for which a licence is requisite under the act, or an Order in Council, or a regulation of a local authority, without having obtained a licence. (2) If, where such a licence is requisite, having obtained a licence,

(*d*) See also the Act, sect. 60, *supra*.

he, with the like intent, does the thing licensed after the licence has expired. (3) If he offers or attempts to use as such a licence an instrument not being a complete licence, or an instrument untruly purporting or appearing to be a licence, unless he shows to the satisfaction of the court of summary jurisdiction before which he is charged that he did not know of that incompleteness or untruth, and that he could not with reasonable diligence have obtained knowledge thereof. (4) If, with intent to unlawfully evade the act, or an Order in Council, or a regulation of a local authority, he alters, or falsely makes, or antedates, or counterfeits, or offers or utters, knowing the same to be altered or falsely made, or antedated, or counterfeited, a licence, declaration, certificate, or instrument made or issued under or for the purpose of the act, or an Order in Council, or a regulation of a local authority: (5) If, for the purpose of obtaining such a licence, certificate, or instrument, he makes a declaration or statement false in any material particular, unless he shows to the satisfaction of the court of summary jurisdiction before which he is charged that he did not know of that falsity, and that he could not with reasonable diligence have obtained knowledge thereof. (6) If he obtains, or endeavours to obtain, such a licence, certificate, or instrument by means of a false pretence, unless he shows to the satisfaction of the court of summary jurisdiction before which he is charged that he did not know of that falsity, and that he could not with reasonable diligence have obtained knowledge thereof. (7) If he grants or issues such a licence, certificate, or instrument, being false in any date or other material particular, unless he shows to the satisfaction of the court of summary jurisdiction before which he is charged that he did not know of that falsity, and that he could not with reasonable diligence have obtained knowledge thereof; or grants or issues such a licence, certificate, or instrument, having, and knowing that he has, no lawful authority to grant or issue the same. (8) If, with intent to unlawfully evade or defeat the act, or an Order in Council, or a regulation of a local authority, he grants or issues an instrument being in form a licence, certificate, or instrument made or issued under the act, or an Order in Council, or a regulation of a local authority, for permitting or regulating the movement of a particular animal, or the doing of any other particular thing, but being issued in blank, that is to say, not being before the issue thereof so filled up as to specify any particular animal or thing. (9) If he uses, or offers or attempts to use, for any purpose of the act, or of an Order in Council, or of a regulation of a local authority, an instrument so issued in blank, unless he shows to the satisfaction of the court of summary jurisdiction before which he is charged that he did not know of its having been so issued in blank, and that he could not with reasonable diligence have obtained knowledge thereof. (10) If, without lawful authority or excuse, proof whereof shall lie on him, he digs up, or causes to be dug up, a carcase buried under the direction of the Privy Council, or of a local authority, or of a receiver of wreck. (11) If, where an Order in Council has prohibited, absolutely or conditionally, the use for the carrying of animals, or for any purpose connected therewith, of a vessel, vehicle, or pen, or other place, he, without lawful authority or excuse, proof whereof shall lie on him, does anything so prohibited. And by sub-sect. 2, in every case in this section specified, he will be liable on conviction to be imprisoned for any term not exceeding two months, with or without hard labour, in lieu of the pecuniary penalty (*e*) to which he is liable under the act.

Other offences.

And by Article 23 of the Rabies Order, 1887, the following are to be deemed offences against the Act of 1878 by the persons mentioned in the first to the fifth sub-section, each according to and in respect of his own acts and defaults, viz.:—(1) If anything is done, or omitted to be done, as regards the muzzling of a dog, or as regards the keeping of a dog under control, in contravention of a regulation of a local authority made under the authority of the order, by the owner of the dog and the person for the time being in charge thereof. (2) If a show or exhibition of a dog or dogs is held in contravention of a regulation of a local authority made under the authority of the order, by the person holding such show or exhibition and the occupier of the place where the show or exhibition is held, and the owner or consignor, or person for the time being in charge of each dog

(*e*) Sect. 60, *ante*, p. 40.

exposed thereat, and the person, if any, taking entrance money or other payment for admission thereto, such last-mentioned person knowing the show or exhibition to be held in contravention as aforesaid. (3) If a dog, &c., or the carcase of a dog, &c., is moved in contravention of a regulation of a local authority made under the authority of the order, or of the conditions of a movement licensed thereunder, it will be deemed an offence by the following, namely, the owner of such dog, &c.; the person for the time being in charge thereof; the person causing, directing, or permitting the movement; the person moving or conveying such dog, &c., or carcase; the owner and the charterer, and the master of the vessel in which it is moved; and the consignee or other person receiving or keeping it, knowing it to have been moved in contravention as aforesaid. (4) It will also be deemed an offence by any person who, with a view to unlawfully evade or defeat the operation of the order, by clipping, or washing, or in any other manner takes out, effaces, or obliterates, or attempts to take out, efface, or obliterate, any mark clipped, painted, or stamped on a dog,

&c., as required by a regulation of a local authority, and by the person causing, directing, or permitting the same to be done, and by the owner of the dog, &c., and by the person for the time being in charge thereof. (5) If anything is omitted to be done, as regards cleansing or disinfection, in contravention of a regulation of a local authority made under the authority of this order, by the owner and the lessee and the occupier of any place or thing in respect of which—and by the person using the van, cart, or other vehicle in which—and by the owner and the charterer and the master of the vessel in which (as the case may be) the same is omitted. (6) If a person in charge of dogs, &c., being moved, where, under a regulation of a local authority, made under the authority of the order, a movement licence is necessary, on demand made under the order, fails to give his true name and address, or gives a false name or address, he will also be deemed guilty of an offence against the act.

Proceedings in court of summary jurisdiction.

By sect. 63 of the Act of 1878, proceedings and penalties for offences against the act may be taken and recovered, and expenses and other money by the act or an Order of Council made recoverable summarily may be recovered with costs, and summary orders under the act may be made with costs by or before a court of summary jurisdiction, under and according to the act 11 & 12 Vict. c. 43, and any act amending the same (*f*).

Appeal.

And by sect. 64, any person thinking himself aggrieved by the dismissal of a complaint by, or by any determination or adjudication of, a court of summary jurisdiction under the act, may appeal therefrom to the next practicable court of general or quarter sessions for the county or place in which the cause of appeal arises, holden not less than twenty-one days after the decision appealed from, on giving notice within ten days after the decision to the clerk of the court whose decision is appealed from of his intention to appeal, and of the grounds thereof, and to the other part. The appellant must also, within three days after such notice, enter into a recognizance before a justice, with two sufficient sureties, conditioned personally to try the appeal. The court may adjourn the appeal, and make such order thereon as it thinks fit.

(*f*) See also 42 & 43 Vict. c. 49, *post*, p. 121.

Nothing in the Rabies Order, 1887, is to be deemed to affect or interfere with the operation of the Dogs Act, 1871 (*g*), or any local or other Act of Parliament for the same or like purposes (*h*).

Carriers' Dogs on Turnpike Roads.

Any person or persons having the care of any wagon, wain, cart, or other carriage, conveying goods for hire or reward, or for sale, on any turnpike road, must chain or fasten any dog that may be attending him or them on such road to such wagon, &c., on pain of forfeiting any sum not exceeding *twenty shillings* (*i*). In default of payment the offender may be committed to prison for *fourteen* days without hard labour (*j*).

(2.) *As to all parts of England, except the Metropolis.*

The 10 & 11 Vict. c. 89 (Town Police Clauses Act), is now incorporated in sect. 171 of the 38 & 39 Vict. c. 55 (Public Health Act, 1875), and consequently applies to the

(*g*) *Ante*, p. 23.
(*h*) The Order, Article 24.
(*i*) 4 & 5 Geo. 4, c. 96, s. 75.
(*j*) 42 & 43 Vict. c. 49, s. 5, *post*, p. 123.

whole of England, except the Metropolis. By sect. 28 of the former act, a person is rendered liable to a penalty of *forty shillings*, or may be committed to prison for a period *not exceeding fourteen days;* and any constable shall take into custody, without warrant, and forthwith convey before a justice, any person who, within his view—(1) Suffers to be at large any unmuzzled ferocious dog, or urges any dog to attack, worry, or put in fear any horse, or other animal. (2) Every owner of any dog who suffers such dog to go at large, knowing or having reasonable ground for believing it to be in a rabid state, or to have been bitten by any dog or other animal in a rabid state. (3) Every person who, after public notice (*k*) given by any justice directing dogs to be confined on account of suspicion of canine madness, suffers any dog to be at large during the time specified in such notice.

Ferocious and dangerous dogs.

(3.) *The Metropolis only.*

By the 2 & 3 Vict. c. 47, s. 54: Every person who, within the limits of the metropolitan police district, in any thoroughfare

2 & 3 Vict. c. 47.

(*k*) As to notice, see *ante*, p. 25.

or public place, suffers any unmuzzled, ferocious dog to be set at large, or sets any dog to attack, worry, or put in fear, any horse or other animal, is liable to a penalty of *forty shillings*. By sect. 61, any constable is empowered to destroy any dog reasonably suspected to be in a rabid state, or which has been bitten by any other dog in a rabid state, and the owner of any dog who permits it to go at large, after having information or reasonable ground for believing it to be in a rabid state, or to have been bitten by any dog or other animal in a rabid state, is liable to a penalty of *five pounds* (*l*).

Sect. 62. And by sect. 62, the magistrate has power in the metropolitan police district to award compensation not exceeding *ten pounds* to any person aggrieved, in addition to any penalty to which the offender may be liable.

30 & 31 Vict. c. 134, s. 18. Under the 30 & 31 Vict. c. 134, s. 18: The police may take possession of any dog found in any street in the metropolis, and not under the control (*m*) of any person, and

(*l*) In default of payment the offender may be sent to prison for one calendar month, the imprisonment to cease on payment of the sum due. The Act, sect. 77.

(*m*) See *ante*, p. 26.

may detain such dog until the owner has claimed the same, and paid all expenses incurred by reason of such detention. The commissioner of police may, if he thinks fit, issue a notice requiring any dog, while in the streets, and not led by some person, to be muzzled in such a manner as will admit of the animal breathing and drinking without obstruction; and the police may take possession of any dog found in the streets without such muzzle during the currency of such order, and may detain such dog until the owner has claimed it, provided a proper muzzle, and paid all expenses connected with its detention. Where any dog taken possession of by the police wears a collar with the address of any person inscribed thereon, a letter stating the fact of such dog having been taken possession of shall be sent by post to the address inscribed on the collar. The commissioner of police may cause any dog which has remained in the possession of the police for three *clear* days, without the owner paying all expenses incurred by its detention, to be sold or destroyed. Where, upon complaint that any dog has bitten, or attempted to bite, any person within the

Police may take possession of stray dogs.

When magistrate may order dog to be destroyed.

metropolis, it appears to the magistrate, having cognizance of such complaint, that such dog ought to be destroyed, the magistrate may so direct, and any police constable may destroy the same accordingly. All dogs detained by the police shall be properly fed and maintained.

Municipal corporations.

It remains for us to observe that bye-laws relating to dangerous dogs have also been made in different boroughs in pursuance of the 5 & 6 Will. 4, c. 76, now repealed by the 45 & 46 Vict. c. 50 (*n*).

(*n*) 45 & 46 Vict. c. 50, s. 219, *post*, p. 81.

CHAPTER IV.

TRESPASSING DOGS—GAME ACTS—COURSING—HUNTING.

Dog spoiling crops.

WHERE a dog trespasses upon land and commits damage, as by spoiling or injuring the crops, or destroying game, no action will be maintainable by the occupier, unless he can show that the dog is of a peculiarly mischievous disposition within the knowledge of his master (*o*). But where a person voluntarily enters upon private or enclosed property accompanied by a dog, he is himself a trespasser, and will be answerable for whatever damage the dog may commit (*p*).

Dog accompanied by master.

Where a dog jumps into a field neither with the consent of, nor by the incitement by

(*o*) *Brown* v. *Giles*, 1 C. & P. 118; *Read* v. *Edwards*, 17 C. B. (N. S.) 245; 11 L. T. (N. S.) 311; 34 L. J. C. P. 31; and see *ante*, p. 4.

(*p*) *Beckwith* v. *Shordike*, 4 Burr. 2093.

his master, it is an involuntary trespass for which no action will lie (*q*).

So where in an action for trespass it appeared that the defendant was a qualified person, and had permission to sport on the right hand side of a road, but his dog crossed over the road, and ranged over a field on the other side, the judge left it to the jury whether the dog had crossed over by the incitement of the defendant or not, and directed them that if they found that the dog escaped against the will of the defendant, and that he had no intention of sporting on the other side of the road, they must find a verdict for him (*r*).

Chasing trespassing beasts with dog.

A man has a right to use a dog for the purpose of driving trespassing beasts off his land, provided he does no more than is absolutely necessary to effect his object. Where the defendant with a little dog chased the plaintiff's sheep out of his grounds, where they were trespassing, and the dog pursued them on to the land of another man adjoining, but could not be recalled at once,

(*q*) *Idem; Brown* v. *Giles*, *supra*; *Read* v. *Edwards*, *supra*.
(*r*) *Dimmock* v. *Allenby*, 2 Marsh. 582.

the plaintiff sued the defendant for chasing his sheep, but it was held that the action was not maintainable, as the trespass was involuntary, the defendant not having incited the dog to chase the sheep after they had left his grounds, but having done his best to recall the dog (*s*). But the chasing of them with a mastiff cannot be justified, and for any damage occasioned to them by such a dog the owner will be responsible (*t*).

Right to seize trespassing dogs.

The occupier of land may have a right to seize and detain dogs trespassing upon or doing damage to it, provided such animals are not in the actual possession and use, and under the control of some person. A dog which a man is leading by a string, or which is under the immediate control of the owner, may not be seized, as such a proceeding might lead to a breach of the peace (*u*); but it seems that a dog which is running about trespassing cannot be considered as under control, although accom-

(*s*) *Mitten* v. *Fundrye*, Poph. 161, cited 4 Burr. 2094.

(*t*) *King* v. *Rose*, 1 Freem. 347.

(*u*) *Field* v. *Adames*, 12 A. & E. 649; *Bunch* v. *Kennington*, 1 Q. B. 679.

panied by a person having the care of him (*v*).

Powers of gamekeepers to seize trespassing dogs.

A gamekeeper is authorized to seize for the use of the lord of the manor, lordship, or royalty, all dogs used for killing and taking game within the limits of his master's manor, &c., by any person not authorized to kill game for want of a certificate (*w*); but it seems that the seizure must be made at the time the party is using the dog for that purpose (*x*). And a gamekeeper is not authorized to seize the dogs of another gamekeeper for merely trespassing out of his proper manor (*y*), as the dogs are generally not the keeper's, but those of his master; and it does not seem to be the intent of the statute that the property of the lord of the manor should be put in the power of his keeper to forfeit whenever he might please to exceed his authority under his deputation (*z*).

(*v*) *Bunch* v. *Kennington*, *supra*.

(*w*) 1 & 2 Will. 4, c. 32, s. 13; and see *post*, p. 69.

(*x*) *Wingfield* v. *Stratford*, 1 Wils. 313; *Rex* v. *Gardner*, 2 Stra. 1098.

(*y*) *Rogers* v. *Carter*, 2 Wils. 387.

(*z*) Deacon's Game Laws, pp. 28, 29.

A gamekeeper authorized by his deputation to "seize greyhounds, setting dogs, and ferrets, and to do all things belonging to his office of gamekeeper according to the directions of the Acts of Parliament," is not thereby authorized to seize hounds (*a*).

Gamekeepers in Wales are authorized to seize dogs used to kill game by persons not holding a certificate (*b*). **Gamekeepers in Wales.**

Deerkeepers may seize the dogs of persons entering with them into any forest, chase, or purlieu for the purpose of hunting, coursing, or killing deer (*c*). **Deerkeepers may seize trespassing dogs.**

Sometimes a man may be justified in destroying a trespassing dog, but in determining whether he was so or not regard will be had to the circumstances under which he destroyed the animal; the place where the transaction occurred; and whether the act was necessary for the protection of his pro- **Right to destroy trespassing dog.**

(*a*) *Grant* v. *Hulton*, 1 B. & Ald. 134.

(*b*) 1 & 2 Will. 4, c. 32, s. 15.

(*c*) 24 & 25 Vict. c. 96, s. 16. A curious and interesting account of the ancient forestal law relating to dogs, and the right to hamble and expeditate them, will be found in Manwood's Forest Laws, Chap. XVI.

perty. The character of the dog should also be taken into consideration.

Dogs trespassing in pursuit of wild animals.

Dogs trespassing in pursuit of animals *feræ naturæ* (wild disposition) (*d*), may not ordinarily be destroyed, even though the owner of the dog has received notice that trespassing dogs will be shot. In the case of *Vere* v. *Lord Cawdor*, Lord Ellenborough observed, "A dog does not incur the penalty of death for running after a hare in another man's ground, and if there be any precedent of that sort which outrages all reason and common sense it is of no authority to govern other cases" (*e*). Neither is it any justification to show that the defendant had put up a notice that dogs trespassing on his land would be shot, and Mr. Justice Dampier reprobated the idea that a man was at liberty to shoot another's dog by virtue of such a notice (*f*).

Dog following game.

A gamekeeper has, therefore, no right to destroy a dog for following game, unless it

(*d*) Animals *feræ naturæ*, or beasts and birds of a wild disposition, such as deer, hares, coneys in a warren, pheasants, partridges, &c., as distinguished from those *domitæ naturæ*, or tame, such as horses, sheep, poultry, &c.

(*e*) *Vere* v. *Lord Cawdor*, 11 East, 569.

(*f*) *Corner* v. *Champneys*, 2 Marsh. 584.

be in a preserve, and the game could not otherwise be saved from destruction, when, it seems, the act may be justified (*g*). So, also, if a dog is chasing fowls in a poultry yard (*h*), rabbits in a warren (*i*), sheep in a fold, or deer in a park (*j*). But after the chasing is discontinued and the peril has ceased, the act cannot be justified. Where, therefore, the owner of sheep in a field shot the dog in another field some distance off, it was held, in an action by the owner of the dog, that the defendant was not justified in the act of shooting the animal, as it was not necessary for the protection of his property (*k*). And a man has no right to shoot a dog for attacking or chasing his beasts or poultry while they are trespassing on his neighbour's land.

Chasing fowls, &c.

The setting of dog spears on land is not illegal, if it appears that they are set in a wood for the mere purpose of destroying

Setting dog spears.

(*g*) *Read* v. *Edwards*, 11 L. T. 311; 17 C. B. (N. S.) 245; 34 L. J. C. P. 31.

(*h*) *Janson* v. *Brown*, 1 Camp. 41.

(*i*) *Wadhurst* v. *Damme*, Cro. Jac. 44.

(*j*) *Barrington* v. *Turner*, 3 Lev. 28; *Protheroe* v. *Matthews*, 5 C. & P. 581.

(*k*) *Wells* v. *Head*, 4 C. & P. 568.

trespassing dogs, and not with the intent to destroy human life, or inflict grievous bodily harm on any person (*l*). So where the plaintiff was passing through a close of the defendant's accompanied by his dog, which, on seeing a rabbit cross the footway, started in pursuit of it, and while in pursuit ran upon a dog spear, and was wounded, it was held to be a good answer to an action by the plaintiff that the defendant set and concealed the engine for the purpose of preserving his game, and of disabling and killing dogs that came upon his close, lest they should pursue and destroy the game, whereof the plaintiff had notice; and it seems that the allegation of such notice is unnecessary (*m*). Whoever places upon his land traps baited with meats so strongly scented as to entice his neighbour's dogs, and draw them irresistibly to their destruction, will be answerable in damages for the injuries sustained. So also, if he entices the

Setting scented traps for dogs, &c.

(*l*) 24 & 25 Vict. c. 100, s. 31.

(*m*) *Jordan* v. *Crump*, 8 M. & W. 787; 5 Jur. 113. See also 24 & 25 Vict. c. 100, s. 31, *ante*.

animals from a highway or public path to the traps on his land; for there is no difference between drawing an animal into the trap by means of his instinct and putting him there by main force (*n*).

The placing of poisoned flesh upon land for the purpose of destroying trespassing dogs is illegal (*o*). Placing poisoned flesh on land.

We must now glance slightly at the provisions of the Game Acts so far as they affect the responsibilities of owners of trespassing dogs, and render them liable to be summarily proceeded against. The principal statute is the 1 & 2 Will. 4, c. 32, whereby a person using a dog for the purpose of killing or taking game (*p*) on Sunday or Christmas-day is liable to a penalty not exceeding *five pounds*.—If for the purpose of killing or taking partridges between the 1st of Feb- Game Acts. 1 & 2 Will. 4, c. 32. Section 3. Sporting at unlawful seasons.

(*n*) See remarks of Lord Ellenborough in *Townsend* v. *Wathen*, 9 East, 277.

(*o*) 27 & 28 Vict. c. 115, s. 2, *post*, p. 91.

(*p*) The definition of the word "game" is variously given in the several statutes on the subject, to each of which reference should be made. In the one now under our notice it includes hares, pheasants, partridges, grouse, heath or moor game, black game, and bustards. The Act, s. 2.

ruary and the 1st of September; or pheasants between the 1st of February and the 1st of October; or black game (except in Somerset, Devon, or the New Forest, in Southampton), between the 10th of December and the 20th of August; or in Somerset, Devon, or the New Forest between the 10th of December and the 1st of September; or grouse, called red game, between the 10th of September and the 12th of August; or bustard between the 1st of March and the 1st of September, to a penalty for every head of game not exceeding *one pound*, with costs (*q*).—If for the purpose of killing or taking game by any person not being authorized to do so for want of a game certificate (*r*), a penalty not exceeding *five pounds*, with costs. A conviction for this offence will not exempt the offender from any penalty under any statute relating to game certificates, but is a cumulative penalty (*s*). It would seem that one who goes

Sporting without certificate.

(*q*) The Act, s. 3.
(*r*) As to game certificates, see *post*, p. 68.
(*s*) The Act, s. 23.

in pursuit of game with a dog and a gun on the same day will only incur a single penalty, although several birds or hares may be killed (*t*).

A person not holding a licence to kill game may lawfully join in the sport of coursing with duly licensed persons, provided he does not use his own dogs (*u*). **Unlicensed person may join in sport of coursing.**

A gamekeeper not holding a game certificate, on which the full amount of duty has been paid, is not authorized to use a dog for the purpose of killing or taking game, except within the limits included in his appointment as gamekeeper; and in the event of his doing so he may be proceeded against in the same manner as if he had no game certificate whatever (*v*). **Gamekeeper holding insufficient game certificate.**

Any person trespassing in pursuit of game in the daytime (*w*) is liable to a penalty not exceeding *two pounds* and the costs of the pro- **Trespassing in pursuit of game in the daytime.**

(*t*) *Regina* v. *Lovat*, 7 T. R. 152.

(*u*) *Molton* v. *Rogers*, 4 Esp. 217; *Lewis* v. *Taylor*, 16 East, 49.

(*v*) 27 & 28 Vict. c. 115, s. 6.

(*w*) Day-time means from the commencement of the last hour before sunrise to the end of the last hour before sunset. The Act, s. 34.

secution, and in the case of five or more persons trespassing together, each is liable to a penalty not exceeding *five pounds*, and the costs of the conviction (*x*).

Any person may prove by way of defence any matter which would have been a defence to an action at law for the trespass, but the leave of the occupier will be no defence when game is reserved to the landlord (*y*).

Dead game not within meaning of the Act. Dead game is not within the meaning of this section, so that the mere entering an adjoining cover with a dog and gun to pick up dead pheasants which had been shot and observed to fall two days before is not a trespass thereunder (*z*).

Sending dog from highway into cover. And the mere sending a dog from a highway into an adjoining cover, unaccompanied by any other act, is not a trespass by entering and being upon land within the meaning of the 30th section. The entry must be personal, and not constructive. But where the

(*x*) The Act, s. 30. It was formerly held that this was a joint offence, and that only one penalty could be inflicted. *Rex* v. *Bleasdale*, 4 T. R. 809.

(*y*) The Act, s. 30.

(*z*) *Taunton* v. *Jarvis*, 68 L. T. 37.

defendant, whilst on a highway carrying a gun, by raising his hand to a dog that accompanied him, sent the dog into a cover on one side of the highway, and immediately afterwards a pheasant flew across the highway, at which the defendant fired, it was held that this constituted a trespass in search of game upon land in the occupation of the lord of the manor and the owner of the land on both sides of the highway (*a*).

Section 31. Trespassers in daytime may be requested to quit land and tell names, &c.

"Trespassers in search of game in the daytime may be requested to quit the land, and to tell their names and addresses, and in case of refusal may be arrested and brought before a justice of the peace within *twelve hours*, but if not brought before a justice of the peace within that time must be discharged" (*b*).

Appeal.

Appeal lies to the Court of Quarter Sessions.

Night Poaching Acts.

Persons have been occasionally convicted of using dogs for the purpose of taking game

(*a*) *Reg.* v. *Pratt*, 24 L. J. M. C. 113.
(*b*) 27 & 28 Vict. c. 115, s. 31.

at night, but as there is no mention of a dog in the Night Poaching Acts (*c*) it would appear that such convictions are bad, and have taken place before justices who were under a misapprehension of the law.

23 & 24 Vict. c. 90. Game licences.

By the 23 & 24 Vict. c. 90, a licence (*d*) must be taken out by every person who uses a dog for the purpose of killing or destroying any game, or any woodcock, snipe, quail or landrail, or any coney or deer; and any person who uses a dog for any of those purposes without a proper licence renders himself liable to a penalty of *twenty pounds.*

(*c*) 9 Geo. 4, c. 69; 7 & 8 Vict. c. 29.

(*d*) The duties payable on these licences are excise duties (sect. 3), and now regulated as follows:—

	£	*s.*	*d.*
Licence or certificate taken out after July 31st and before November 1st to expire on July 31st in the following year - - - - - -	3	0	0
To expire on October 31st in the same year in which the licence or certificate shall be taken out - - - - - - - - - -	2	0	0
Licence or certificate taken out on or after November 1st to expire on July 31st following	2	0	0
Licence to be taken out for a gamekeeper - -	2	0	0

The excise duty payable on a licence to carry a "*gun*," which term includes a firearm of any description, and an air-gun, is *ten shillings*, and regulated by the 33 & 34 Vict. c. 57, and 46 & 47 Vict. c. 10, s. 6.

But exceptions are allowed in the case of taking or destroying conies in Great Britain by the proprietor of any warren or enclosed ground; or by the tenant of lands either by himself or by his direction or permission; the pursuing and killing of hares respectively by coursing with greyhounds, or by hunting with beagles or other hounds; the pursuing and killing of deer in enclosed lands by the owner or occupier of such lands, or by his direction or permission. And exemptions are made in favour of any of the royal family; any person appointed a gamekeeper on behalf of her Majesty's woods, forests, and land revenues under the authority of any act of Parliament relating to the lands or revenues of the Crown; any person aiding or assisting in the taking or killing of game, or any woodcock, snipe, quail, landrail, or coney or deer in the company or presence, and for the use of another person, who shall have duly obtained, and in his own right, a licence to kill game, and who shall then and there use his own dog for taking and killing the same, and who shall not act therein by virtue of any

Exceptions.

Exemptions.

deputation or appointment; and as regards the killing of hares only, all persons who, under the 11 & 12 Vict. c. 29, are authorized to kill hares without a certificate (*e*).

Hunting and coursing. The provisions of the 1 & 2 Will. 4, c. 32, against trespassers and persons found on any land, do not extend to any person hunting or coursing with hounds or greyhounds, and being in fresh pursuit of any deer, hare, or fox already started upon any other land (*f*).

Extirpation of vermin It was formerly considered, and so held in several cases, that a person might justify hunt-

(*e*) The persons here indicated are such as are in the actual occupation of enclosed ground, owners having the right to kill game thereon, and persons authorized by them to do so. And by the 43 & 44 Vict. c. 47, s. 1, every occupier of land has, as incident and inseparable from his occupation, the right to kill and take ground game (*i. e.*, hares and rabbits (s. 8)) concurrently with any other person entitled to kill and take ground game on the same land: provided that he shall kill and take ground game only by himself or by persons authorized by him in writing. No person shall be authorized by the occupier to kill or take ground game, except members of his household resident on the land in his occupation; persons in his ordinary service on such land; and any person *bonâ fide* employed by him for reward in the taking and destruction of ground game.

(*f*) S. 35. See also *Paul* v. *Summerhayes*, 4 L. R. Q. B. Div. 9; 48 L. J. M. C. 33.

ing over the grounds of another, provided the governing motive was the extirpation of vermin, and not merely the pleasure of a "run," and that he did no more than was necessary to kill the fox, otter, or other object of the chase (*g*). **governing motive.**

But persons hunting for mere pleasure over a man's land without his leave, have always been regarded as trespassers (*h*). And the owner of hounds will be answerable for the damage which huntsmen may commit by trespassing on the lands of another with the hounds, although he does not personally accompany them, unless it can be shown that they went there contrary to his request (*i*). It has always been considered that even where the object was the extirpation of vermin, huntsmen must do no more than was necessary, and not maliciously **Hunting for pleasure.**

(*g*) *Nicholas* v. *Badger*, 3 T. R. 259, n.; *Gedge* v. *Minue*, 2 Bulst. 62.

(*h*) *Earl of Essex* v. *Capel*, Chitty's Game Laws, 32, n., cited in Selwyn's Nisi Prius, 13th ed. p. 838; *Paul* v. *Summerhayes, supra*.

(*i*) *Baker* v. *Berkeley*, 3 C. & P. 32.

destroy fences or injure crops (*j*). In the case of *Hume* v. *Oldacre* the defendant was held liable for the whole of the damage so occasioned both by himself, and also by the party of huntsmen who accompanied him (*k*).

In the present day it could scarcely be successfully contended that the governing object of the chase was the extirpation of vermin, since hunting is now so well recognized as a sport and means of diversion, to which the motive of destroying noxious animals is at most only collateral. The interest and excitement of the chase is the main object. The sport must be carried on in subordination to the ordinary rights of property, so that a person has no right, in the pursuit of a fox, to come upon the land of another against his will. Those, however, who pursue the sport, generally do so in a reasonable spirit, and

(*j*) *Gundry* v. *Feltham*, 1 T. R. 334. See also *Nicholas* v. *Badger*, 3 T. R. 259; *Gedge* v. *Minue*, 2 Bulstr. 62.

(*k*) *Hume* v. *Oldacre*, 1 Stark. 252. This has been acted upon in more recent unreported cases.

only go upon the lands of those whose consent is expressly or may be assumed to be tacitly given(*l*).

(*l*) See the judgments of Lord Coleridge, C. J., and Mellor, J., in *Paul* v. *Summerhayes*, *supra*.

CHAPTER V.

NUISANCES CAUSED BY DOGS.

Whoever keeps any animal so as to be a nuisance or injurious to health commits an offence under the 38 & 39 Vict. c. 55 (Public Health Act, 1875) (*m*). Information may be given to the local authority by any person aggrieved, or by any two householders, or any officer of such authority, or the relieving officer, or any constable or officer of the police force of the district (*n*).

On receipt of information the local authority may, if satisfied of the existence of the nuisance, serve a notice on the offender requiring him to abate it within a time specified in the notice (*o*). If default is made in complying with it, the local authority must cause a complaint to be made to a justice, who will thereupon issue a

(*m*) 38 & 39 Vict. c. 55, s. 91.
(*n*) *Idem*, s. 93.
(*o*) *Idem*, s. 94.

summons requiring the offender to appear before a court of summary jurisdiction (*p*); when, if satisfied that the nuisance exists, it will make an order for its abatement, and may also impose a penalty not exceeding *five pounds*, and give directions as to the payment of all costs incurred up to the hearing or making the order for abatement (*q*).

Notices, orders, and other documents, may be served by delivering the same at the residence of the person to whom they are addressed, or, when addressed to the owner or occupier of the premises, by delivering the same, or a true copy thereof, to some person on the premises; or, if there be no such person, by fixing the same on some conspicuous part of the premises, or by post by a prepaid letter, and, if served by post, it will be deemed to have been served at the time when the letter containing the same would be delivered in the ordinary course of post; and, in proving such service, it will be sufficient to prove that the notice, order or other document, was properly addressed and

(*p*) *Idem*, s. 95.
(*q*) *Idem*, s. 96.

put into the post, and notices may be addressed to the "owner" or "occupier" of the premises (*r*).

An appeal lies from the decision of a court of summary jurisdiction to the court of quarter sessions (*s*).

Boroughs. The offence is also provided for in the bye-laws of certain boroughs (*t*).

Nuisance caused by barking. It is said, that when a dog creates a nuisance by barking, and so disturbing the neighbours, an action may be brought against the owner of the animal for damages, and that the court may grant an injunction against a continuance of the nuisance. But in an action against the defendant for keeping dogs so near the plaintiff's dwelling-house that he was disturbed in the enjoyment thereof, it appeared in evidence that the defendant kept six or seven pointers so near the plaintiff's dwelling-house that his family were prevented from sleeping during the night, and were very much disturbed in the day-time. No evidence was given for the defendant, notwithstanding which the jury

(*r*) The Act, s. 267.
(*s*) *Idem*, s. 269.
(*t*) *Post*, p. 81.

found a verdict for him, and on a motion for a new trial it was refused by the court (*u*). And in the recent case of *Huddlestone* v. *Boare* (*v*), the plaintiff, who was a gentleman devoted to scientific pursuits, and residing at Weybridge, Surrey, applied for an injunction to restrain the defendant, who occupied the adjoining premises, from keeping a dog so as to create a nuisance by barking. It was stated that the defendant kept the dog, which barked and yelped so much during the day as to cause the plaintiff the greatest annoyance; that it was an intolerable nuisance to his household, depriving them of their proper rest, and seriously interfering with his scientific pursuits; and that, having erected a laboratory at great expense, he had been unable to use it for some time in consequence of the noise occasioned by the animal. Mr. Justice Pearson said he was very sorry he could not grant any relief to the plaintiff from the evils complained of, although he could quite understand how a gentleman devoted to science should feel even ordinary noises in his immediate neigh-

Huddlestone v. *Boare.*

(*u*) *Street* v. *Tugwell*, Queen's Bench, M. T. 41 Geo. 3, M. S.
(*v*) *Huddlestone* v. *Boare*, Ch. Div. July 22nd, 1886.

bourhood as intolerable whilst pursuing his studies. His Lordship added that it could not be expected that a person residing in a semi-rural district should be as free from the noise complained of as if he were living in a large city. Whilst admitting that a gentleman engaged in such pursuits, as was the plaintiff, might feel the barking of the dog in his neighbourhood as a great annoyance, he thought it highly probable that he would equally suffer, if residing in any other part of the country, if he had a neighbour who kept a watch-dog on the premises . . . . . It appeared to him that the witnesses for the plaintiff might, though unconsciously, have represented the noise from the barking of the dog as being greater than it really was.

The result of these cases can scarcely be deemed satisfactory. We must bear in mind that no action can be maintained for the *reasonable* use of a man's right, even though it may tend to the inconvenience of another; also that a man has a right to keep a dog for the protection of his premises, whose barking would be most likely to scare thieves. But it is greatly to be feared that the noise thus occasioned is too frequently productive of illness and injurious effects on invalids. If a plaintiff could show that such consequences had arisen, it would probably be held a sufficient ground for the action; but it would seem that, unless a very strong case can be made out, the court will not interfere.

Proceedings may be taken by way of indictment for a public nuisance, in which case the nuisance should be described according to the circumstances; and it should be stated to be continuing, if that be the fact, otherwise there cannot be judgment for abatement (*w*).

Summary proceedings.

Whether courts of summary jurisdiction have power to order an abatement of the noise caused by the barking of a dog as a nuisance is very doubtful (*x*), and the opinions of the London stipendiary magistrates are at variance with one another on this point. Occasionally orders for abatement have been made by them, but as a rule applicants are referred to the county court.

In a recent case a summons was taken out under the 18 & 19 Vict. c. 121 (Nuisances Removal Act), against the defendant for

(*w*) *Rex* v. *Stead*, 8 T. R. 142.
(*x*) But see *post*, p. 81.

keeping dogs so as to be a nuisance and injurious to the health of the neighbours. The magistrate of the Clerkenwell Police Court (Mr. Hosack) made an order for the defendant to abate the nuisance forthwith, and to pay the costs of the summons. But on the following day a gentleman applied to the magistrate at the Wandsworth Police Court (Mr. Bennett) for his advice on behalf of himself and his neighbours. He stated that they were suffering from the perpetual howling of a dog, and referred to the above-mentioned case, decided by Mr. Hosack on the previous day. Mr. Bennett said the applicant would have to prove that the howling of the dog was a nuisance, and injurious to health; to which the applicant replied, that he thought it was very prejudicial to health to have a dog perpetually howling. Mr. Bennett said his own impression was that the act did not refer to dogs, but to cattle; the applicant could apply again, and in the meantime consider whether he would take out a summons or proceed by indictment at common law (*y*).

(*y*) Daily News, Feb. 9th, 1887.

This offence is also provided for in the bye-laws of some boroughs (*z*).

It is perhaps needless to say that it is very desirable in all such cases as the above that the parties should endeavour to meet and reconcile their differences without resorting to a court of law.

Upon the nuisance question there have been many County Court decisions where the crowing of cocks has been objected to. In these cases the judges have made what seems to be a reasonable distinction between keeping a cock which crows in a town and in a rural district, and have held that what might properly be held to be a nuisance to the inhabitant of a London street or square might be considered to be lawful in the country.

(*z*) The council of any borough may make bye-laws for its good rule and government, and for the prevention and suppression of nuisances not already punishable in a summary manner by virtue of any act in force throughout the borough, and may appoint fines, not exceeding *five pounds*, for the prevention and suppression of offences against the same. 45 & 46 Vict. c. 50 (repealing 5 & 6 Will. 4, c. 76), s. 23. Any offence against such a bye-law may be prosecuted summarily, on information being laid within six calendar months from the commission of the offence. An appeal lies to the court of quarter sessions. *Idem*, s. 219.

CHAPTER VI.

INJURIES TO DOGS—CRUELTY TO ANIMALS.

HITHERTO we have been chiefly considering the duties and responsibilities of those who keep dogs. We shall now endeavour to explain what rights such persons possess in respect of the animals; the remedies and modes of proceeding for the destruction of, or injury to them; also how the owners may themselves incur liability under the statutes for the prevention of cruelty to animals.

Destroying dogs, when justifiable or not.

An action may be maintained for the destruction of a dog, and the circumstance of its being of a ferocious disposition, and at large, is not sufficient to justify a man in shooting it. But if, when loose, it is actually attacking him, he may destroy it at the time in self-defence, though not after the peril has ceased (*a*). Where the defendant was passing the plaintiff's house, and the

(*a*) *Clark* v. *Webster*, 1 C. & P. 104; *Hannay* v. *Boultbee*, 4 C. & P. 350; 1 M. & R. 15.

plaintiff's dog ran out, and bit the defendant's gaiter, and on the defendant turning round and raising his gun the dog ran away, and the defendant then shot the dog as it was running away, it was held that he was not justified in so doing (*b*).

The shooting of a dog for worrying trespassing sheep or fowls cannot be justified (*c*). And, as we have already seen, a gamekeeper has no right to shoot a dog for following game (*d*), although the owner has received notice that trespassing dogs will be shot (*e*).

Evidence may be given as to general quietness of dog in action for shooting it.

If an action is brought for the shooting of a dog, the plaintiff may call witnesses to prove its general quietness (*e*). On the other hand, the defendant may produce evidence as to the mischievous propensities of the animal which may be considered in mitigation of damages (*f*).

Use of dog for purpose of

The use of a dog for the purpose of drawing, or helping to draw, any cart,

(*b*) *Morris* v. *Nugent*, 7 C. & P. 572.
(*c*) *Wells* v. *Head*, 4 C. & P. 518, *ante*, p. 61.
(*d*) *Vere* v. *Lord Cawdor*, 11 East, 569, *ante*, p. 60.
(*e*) *Idem.*
(*f*) *Wells* v. *Head*, 4 C. & P. 518, *supra*.

draught prohibited. carriage, truck, or barrow is prohibited in every part of the United Kingdom under a penalty of not more than *forty shillings* for the first offence, and not more than *five pounds* for the second or subsequent offence (*g*), and in default of payment the offender may be sent to prison for a term not exceeding one calendar month, the imprisonment to cease on payment of the sum due (*h*).

12 & 13 Vict. c. 92 (Prevention of Cruelty to Animals). By the 12 & 13 Vict. c. 92, s. 2, "If any person shall cruelly beat, ill-treat, over-drive, abuse, or torture, or cause or procure to be cruelly beaten, &c., any animal (*i*), every such offender shall, for every such offence, forfeit and pay a penalty not exceeding *five pounds*."

Performing operation which causes pain. It has been held that to perform an operation upon an animal which causes pain is to "cruelly ill-treat, abuse, or torture the animal" within the meaning of this section, unless the act is justified by showing that it was done for some useful purpose, legalized by custom, for the benefit of the animal

(*g*) 2 & 3 Vict. c. 47, s. 56, amended by 17 & 18 Vict. c. 60, s. 2.

(*h*) 2 & 3 Vict. c. 47, s. 77.

(*i*) The word animal includes "dog." The Act, s. 29.

itself, or for making it more serviceable for the lawful use of man (*j*).

But, to constitute an offence, the cruelty must be of an active kind; mere passive cruelty not being contemplated in the statute. The case of *Powell* v. *Knights* (*k*) came before the Queen's Bench Division of the High Court upon a case stated for the opinion of the court by a stipendiary magistrate. The appellant had shot with a bullet from a pistol, through the eye, a neighbour's dog, which was trespassing in his garden, and the magistrate assumed for the purpose of the case that the shooting was lawful. The appellant immediately gave notice of what he had done to the owner of the dog, and to a policeman, and dragged the dog into the road. When in the road he found, for the first time, that the dog was not dead, but was in pain. He, however, left it in the road, and, though desired to do so by the policeman, refused to cause it to be killed. It died some hours afterwards in great pain, and the appellant knew when he left it that it could not survive

Powell v. *Knights.*

(*j*) *Murphy* v. *Manning*, 36 L. T. (N. S.) 592; 25 W. R. 540; 46 L. J. 211; 2 L. R. Ex. Div. 397. And see 39 & 40 Vict. c. 77 (Vivisection Act), *post*, p. 92.

(*k*) *Powell* v. *Knights*, 38 L. J. (N. S.) 607; 26 W. R. 721.

its wound, but that death only could put an end to its pain. The court, consisting of Cockburn, L. C. J., and Mellor, J., while strongly condemning the inhuman conduct of the appellant, held, nevertheless, that he had not been guilty of any offence within the act, and that the conviction of the magistrate was wrong (*l*).

Section 3. Fighting, baiting, &c., animals.

By sect. 3, "Every person who shall keep or use or act in the management of any place for the purpose of fighting or baiting any bull, bear, badger, *dog*, cock, or other kind of animal, whether of a domestic or wild nature, or shall permit or suffer any place to be so used, shall be liable to a penalty not exceeding *five pounds* for every day he shall so keep or use or act in the management of any such place, or permit or suffer any place to be used as aforesaid: Provided always, that every person who shall receive money for the admission of any other person to any place kept or used for any of the purposes aforesaid, shall be deemed to be the keeper thereof; and every person who shall in any manner encourage, aid, or assist at the fight-

(*l*) See also *Everett* v. *Davies*, 38 L. T. (N. S.) 360; 26 W. R. 332.

ing or baiting of any bull, bear, badger, *dog*, cock, or other animal as aforesaid, shall forfeit and pay a penalty not exceeding *five pounds* for every such offence" (*m*).

It is no offence under this section to encourage, aid, &c., except in a place kept or used for that purpose, but it seems that it is such an act of cruelty as is contemplated by sect. 2 (*n*). Where a match took place in a field containing an area of three acres, walled in so that the rabbits could not escape, between the owners of two dogs, as to which could kill the greatest number of rabbits by running after them, it was held that this was not baiting animals within the meaning of this section (*o*).

Section 4. Liability for injury caused by beating, &c.

And, by sect. 4, "If any person shall by cruelly beating, ill-treating, over-driving, abusing, or torturing any animal, do any damage or injury to such animal, or shall thereby cause any damage or injury to be

(*m*) Somewhat similar provisions are contained in the 2 & 3 Vict. c. 47, s. 47; and the 10 & 11 Vict. c. 89, s. 36 (incorporated in sect. 171 of the 38 & 39 Vict. c. 55, Public Health Act, 1875).

(*n*) *Clark* v. *Hayne*, 29 L. J. M. C. 105; 2 L. T. R. (N. S.) 85; *Morley* v. *Greenhalgh*, 32 L. J. M. C. 93.

(*o*) *Pitts* v. *Millar*, 9 L. R. Q. B. 380; 43 L. J. M. C. 96; 30 L. T. (N. S.) 328.

done to any person, or to any property, every such offender shall, on conviction of such offence, pay to the owner of such animal (if the offender shall not be the owner thereof), or to the person who shall sustain damage or injury as aforesaid, such sum of money by way of compensation, not exceeding the sum of *ten pounds*, as shall be ascertained and determined by the justice of the peace by whom such person shall have been convicted: Provided always, that the payment of such compensation, or any imprisonment (*p*) for the non-payment thereof, shall not prevent, or in any manner affect, the punishment to which such person, or the owner of such animal, may be liable for or in respect of the beating, ill-treating, or abusing of the said animal: Provided also, that nothing herein contained shall prevent any proceeding by action against such offender, or the employer of such offender, where the amount, or damage, or injury is not sought to be recovered under this act.

Section 12. Improperly conveying animals.

By sect. 12, "Any person who conveys or carries, or causes to be conveyed or carried, in or upon any vehicle, any animal in such a

(*p*) See sect. 18, *post*.

position as to subject such animal to unnecessary pain or suffering, is liable to a penalty not exceeding *three pounds* for the first offence, and a penalty of *five pounds* for the second and every subsequent offence."

By sect. 14, "A complaint under this act must be made within one calendar month after the offence is committed." Section 14. When complaint must be made.

By sect. 18, "In case of a conviction, and on default being made in payment of the penalty imposed, or the amount awarded as compensation or damage, together with the costs (if any), the offender may be imprisoned, with or without hard labour, for any time not exceeding two calendar months, unless payment be sooner made. And where the conviction takes place before two justices, or before a metropolitan police magistrate, the offender may be committed to prison, with or without hard labour, for any time not exceeding three calendar months without the option of paying a pecuniary penalty." Section 18. Offenders not paying penalty may be committed.

By sect. 25, an appeal lies to the next court of general or quarter sessions, where the sum adjudged to be paid exceeds *two pounds*, or where imprisonment is adjudged. Section 25. Appeal.

The unlawful killing, maiming, or wound-

24 & 25 Vict. c. 97. Section 41. Killing, maiming, &c., animals.

ing a dog is an offence punishable, on summary conviction, by imprisonment, with or without hard labour, for any term not exceeding *six calendar* months (*q*), or else by forfeiture, over and above the amount of injury done, of a sum of money not exceeding twenty pounds. A subsequent offence is punishable by imprisonment with hard labour for a term not exceeding twelve months (*r*). But, under the Summary Jurisdiction Act, 1879, if the conviction is by one justice, the imprisonment must not exceed *fourteen days*, and the sum adjudged to be paid must not exceed twenty shillings (*s*).

Section 66. Justice may discharge offender in case of first conviction.

In the case of a first conviction, the justice may, if he think fit, discharge the offender, upon his making satisfaction to the party aggrieved for damages and costs, or either of them (*t*).

Section 68. Appeal.

An appeal lies to the next court of general or quarter sessions, where the sum adjudged to be paid exceeds *five pounds*, or the imprison-

(*q*) On appearing before the court, and before the charge is gone into, the defendant may claim to be tried by a jury. 42 & 43 Vict. c. 49, s. 17, *post*, p. 124.

(*r*) 24 & 25 Vict. c. 97, s. 41.

(*s*) 42 & 43 Vict. c. 49, s. 20.

(*t*) 24 & 25 Vict. c. 97, s. 66.

ment adjudged exceeds one month, or the conviction takes place before one justice only (*u*).

Placing poisoned flesh on land.

The placing of poisoned flesh in an inclosed garden for the purpose of destroying trespassing dogs, is not an offence punishable under the 41st section; but it seems that it is within the 27 & 28 Vict. c. 115, s. 2, which renders a person placing poisoned meat upon any land liable to a penalty of *ten pounds* (*v*), or, on default in payment, to imprisonment, with or without hard labour, for a term not exceeding *two calendar months* (*w*); but the occupier of any dwelling-house, or other building, or the owner of any rick or stack of wheat, &c., or of any cultivated garden produce, may lawfully place poison or poisonous ingredients in such dwelling-house or building, or in any inclosed garden attached to such dwelling-house, or in the drains connected with it, for the purpose of destroying rats, mice, or other small vermin, provided that such drains are protected with gratings, or otherwise, so as to prevent any dog from

(*u*) *Idem*, s. 68.

(*v*) *Daniel* v. *Janes*, 2 L. R. C. P. Div. 351.

(*w*) See *ante*, p. 63.

entering the same, or within any rick or stack(v).

39 & 40 Vict. c. 13. Section 1. Administering poisonous drug.

Any person administering to, or causing to be administered to or taken by, any domestic animal any poisonous or injurious drug or substance (unless some reasonable cause or excuse is shown), is liable to a penalty not exceeding the sum of *five pounds*, or to imprisonment, with or without hard labour, for a term not exceeding *one month*, in the case of a *first offence*, or *three months* in the case of a second or subsequent offence(x); but the act is not to exempt the offender from greater or other punishment under any other act or law, so that he be not more than once punished for the same offence(y). The act does not extend to the owner or person acting by the authority of the owner of the animal(z).

39 & 40 Vict. c. 77 (Vivisection). Who may perform experiments on living animals.

By the 39 & 40 Vict. c. 77 (Vivisection Act, 1876)(a), a person duly qualified may, on obtaining a licence from the Secretary of State, and subject to certain restrictions, perform experiments on living animals, in a

(v) The Act, s. 3.
(x) 39 & 40 Vict. c. 13, s. 1.
(y) *Idem*, s. 3.
(z) *Idem*, s. 2.
(a) This act will be found in the Appendix, p. 137.

registered place, with a view to the advancement by new discovery of physiological knowledge, or of knowledge which will be useful for saving or prolonging life, or alleviating suffering, or for the purpose of testing a particular former discovery alleged to have been for the advancement of such knowledge (*b*).

Experiments on dogs and cats.

An experiment calculated to give pain may not be performed on dogs and cats, except on a certificate stating, in addition to the other statements required, that for the reasons specified in the certificate, the object of the experiment will be necessarily frustrated, unless it is performed on an animal similar in construction and habits to a cat or dog, and no other animal is available for such experiment (*c*).

Penalty for performing experiment in contravention of act.

Any person performing, or taking part in performing, any experiment calculated to give pain, in contravention of the act, is, for a first offence, liable to a penalty not exceeding *fifty pounds*, and for a second or any subsequent offence to a penalty not exceeding *one hundred pounds*, or to imprisonment for a period not exceeding *three months* (*d*).

(*b*) The Act, s. 3.
(*c*) *Idem*, s. 5.
(*d*) *Idem*, s. 2.

Exhibition to the public.

Any exhibition to the general public, whether admitted on payment of money or gratuitously, of experiments on living animals, calculated to give pain, is illegal; and any person performing, or aiding in performing, such experiments, is, for a first offence, liable to a penalty not exceeding *fifty pounds*, and for a second or any subsequent offence to a penalty not exceeding *one hundred pounds*, or to imprisonment for a period not exceeding *three months*; and any person advertising the exhibition by newspaper, placard, or otherwise, is liable to a penalty not exceeding *one pound* (*e*). A person punished for an offence under the 6th section of the act is not for the same offence punishable under any other section of the act (*f*).

Judge may grant licence for performance of experiments in criminal case.

A judge of the High Court of Justice may grant a licence for the performance of experiments when necessary in a criminal case (*g*).

Justice may issue warrant.

A justice of the peace, on information on oath that there is reasonable ground to believe that experiments in contravention of the act are being performed by an unlicensed person

(*e*) In default of payment, fourteen days' imprisonment. 42 & 43 Vict. c. 49, s. 5. See *post*, p. 123.

(*f*) 39 & 40 Vict. c. 77, s. 6.

(*g*) *Idem*, s. 12.

in any place not registered under the act, may issue his warrant, authorizing any officer or constable of police to enter and search such place, and to take the names and addresses of the persons found therein, and any person who refuses admission on demand to a police officer or constable, or who refuses to disclose his name and address, is liable to a penalty not exceeding *five pounds* (*h*), or in default in payment to imprisonment for a term not exceeding *two months* (*i*).

Right to trial by indictment. **Appeal.**

Where the penalty exceeds five pounds the accused may elect to be tried by indictment (*j*); and any party thinking himself aggrieved by any conviction made by a court of summary jurisdiction may appeal therefrom to the court of quarter sessions (*k*).

Prosecution not to be instituted without consent of Secretary of State.

A prosecution under the act against a licensed person is not to be instituted, except with the consent in writing of the Secretary of State.

(*h*) *Idem*, s. 13.
(*i*) 42 & 43 Vict. c. 49, s. 5, *post*, p. 123.
(*j*) 39 & 40 Vict. c. 77, s. 16.
(*k*) *Idem*, s. 21.

CHAPTER VII.

LIABILITY OF CARRIERS AND RAILWAY COMPANIES FOR NEGLIGENCE IN CONVEYING DOGS—DOG SUBJECT OF PAWN.

A COMMON carrier is one who undertakes for a pecuniary consideration to convey goods from place to place, either by land or water. Being in the nature of an insurer (*l*), his warranty is to carry them safely and securely. He is responsible for the loss of them by any cause except the negligence of the owner, or an inherent defect, or act of God (*m*), or of the enemies of the Queen. Railway companies are not common carriers of dogs (*n*), but may restrict their liability by

(*l*) *Aliter*, as to railway companies, see *post*, p. 103.

(*m*) By the term "act of God" is meant something independent of the will of man, such as storms, gusts of wind, lightning, inundations, sudden death or illness, and inevitable accidents not resulting from human agency.

(*n*) *Dixon* v. *Great Northern Ry. Co.*, 3 Times Law Reports, 200.

means of a special contract, which must, however, be just and reasonable, and signed by the customer or person who delivers the goods for carriage, or he will not be bound thereby (*o*). If they carry a dog without making a special contract, they will be liable as common carriers in the ordinary way (*p*).

A dog with a string round his neck having been delivered to a common carrier to be carried, and tied by the string in a watch box, shortly after slipped his head through the noose, and escaped. In an action to recover the value of the dog, it was contended that the owner ought to have taken care that the string was properly secured round the dog's neck; but it was held that, as the carrier had the means of seeing the dog was insufficiently secured, he ought to have locked the dog up, or taken other proper means to secure the animal, and that he was responsible for the loss (*q*). But where a greyhound, secured in the ordinary

(*o*) *Harrison* v. *London, Brighton & South Coast Ry. Co.* (Ex. Ch.), 31 L. J. Rep. (N. S.) Q. B. 113; 2 Best & S. 122; *Dixon* v. *Great Northern Ry. Co.*, *supra.*

(*p*) *Idem.*

(*q*) *Stuart* v. *Crawley*, 2 Stark. 324.

way, and according to the obvious intention of the owner, viz., by a collar and strap, was delivered to a railway company to be carried, and the greyhound, during the journey, slipped his head through the collar, and was lost, it was held that the company were not responsible (*r*).

17 & 18 Vict. c. 31 (Railway and Canal Traffic Act).

Under the 17 & 18 Vict. c. 31, s. 7 (Railway and Canal Traffic Act, 1854), railway and canal companies are liable for the loss of or injury to animals in the receiving, forwarding, or delivering thereof, occasioned by the default of such companies or their servants, notwithstanding any notice, condition, or declaration, made and given by such companies contrary thereto, or in anywise limiting such liability; every such notice, condition, or declaration being thereby declared to be null and void. Provided that nothing therein contained shall be construed to prevent the companies from making such conditions with respect to the receiving, forwarding and delivering such animals, &c., as shall be adjudged by the court or judge, before whom any question relating thereto shall be tried,

(*r*) *Richardson* v. *North Eastern Ry. Co.*, L. R. 7 C. P. 75.

to be just and reasonable. Provided also, that no special contract between such companies and any other parties respecting the receiving, forwarding, or delivering of any animals, &c., shall be binding upon or affect any such party, unless the same shall be signed by him or by the person delivering such animals, &c., for carriage.

The plaintiff delivered to the defendants, a railway company, a dog to be carried, and signed a ticket containing the following terms: "Received the annexed ticket, subject to the following conditions: The company will not be liable in any case for loss or damage to any horse or other animal above the value of 40*l.*, or of any dog above the value of 5*l.*, unless a declaration of its value, signed by the owner or his agent at the time of booking, shall have been given to them; and by such declaration the owner shall be bound, the company not being in any event liable to any greater amount than the value declared. The company will in no case be liable for injury to any horse or other animal, or dog, of whatever value, where such injury arises wholly or partially from fear or res-

F 2

tiveness. If the declared value of any horse or other animal exceed 40*l.*, or any dog 5*l.*, the price of conveyance will, in addition to the regular fare, be at the rate of 2*l.* 10*s.* per cent. upon the declared value above 40*l.*, whatever may be the amount of such value, and for whatever distance the animal is to be carried." The value of the dog was 21*l.*, but the plaintiff made no declaration of its value, and paid only the regular fare, 3*s.* The dog escaped from the train during the journey, without any neglect or default on the part of the company. The plaintiff having sued the company for the loss, it was held in the Exchequer Chamber (*dissentiente* Wilde, B.), reversing the judgment below, that the plaintiff was not entitled to recover; Erle, C. J., and Keating, J., being of opinion that sect. 7 (*s*) of the Railway and Canal Traffic Act, 1854 (17 & 18 Vict. c. 31), was confined in its application to cases where the loss or injury was occasioned by the neglect or default of the company, and had no bearing on a case such as the present, where the loss arose from pure accident, and that the

Harrison v. London & South Coast Railway Co.

(*s*) *Supra.*

company were exempt from liability by the terms of the contract. And it was held, further, that assuming that the statute applied to this case, the conditions in the ticket were reasonable and just, and that they were not to be construed as meaning to exempt, or as having the effect of exempting, the company from liability for loss or injury occasioned by wilful misconduct on their part (*t*).

Dixon v. Great Northern Railway Co.

But in the recent case of *Dixon* v. *The Great Northern Railway* (*u*), which came before the court on appeal from the Queen's Bench Division, the plaintiff sought to recover damages from the defendants for injuries caused to his coursing dog, "Dutch Oven," by reason of the alleged negligence of the defendants' servants. It appeared that the dog in question was entered for the Gosforth coursing stakes for 1,000*l.*, and that it was sent from King's Cross to Newcastle. The animal arrived safely at Newcastle, but while it was in the central station there, a porter

(*t*) *Harrison* v. *London, Brighton & South Coast Ry. Co.*, *supra.*

(*u*) *Supra.*

in the defendants' service ran the wheel of a truck over the dog's tail and broke it, the wound being of such a character as to prevent it from coursing. The man in charge of the dog at King's Cross station signed his name to a ticket, which stated that the company would not become common carriers of dogs, and that they would not convey such animals except on the terms that they would not be responsible for any injury that might be done to them, beyond the sum of 2*l.*, unless the higher value was declared, and 5 per cent. were paid on the higher value. A county court judge had held that the terms imposed by this ticket were unreasonable (*v*), but his decision had been reversed by the Divisional Court. On the appeal, the Master of the Rolls (Lord Esher) said: "It is quite clear that railway companies are not common carriers of dogs, and, therefore, they are not liable in this case as common carriers. They cannot, at common law, be forced to carry dogs at all, and they may make any terms they please

(*v*) See *ante*, p. 97.

with any individual, who wishes them to carry his dog. But whether they are common carriers or not, if they had carried a dog for money, and made no terms regarding it, it was beyond doubt that they would subject themselves to the ordinary liabilities of bailees. Whether, however, the company are liable here, depends upon the terms of the Railway and Canal Traffic Regulation Acts. What we have to determine, therefore, is, whether the condition which the company have imposed is just and reasonable. That condition, if enforced, would have absolved the company for any injury caused by wilful negligence or dishonesty. I think such a condition unreasonable, but a superior authority has said that a company under such circumstances would not be liable, provided they had given a reasonable alternative to the owner of the animal which was to be sent by their line. In the present case the ordinary fare for the conveyance of the dog was 6*s.*, but under the conditions stated on the ticket it would have been 3*l.* 4*s.* As the company have no power to act as insurers, I regard that charge as an unreasonable one, and as not offering a reasonable alternative to the sender. I am, therefore, of opinion that the decision of the Divisional Court was wrong, and should be reversed, and that judgment should be entered for the plaintiff with costs." Lindley and Lopes, L.JJ., concurred.

Dog may be subject of pawn.

A dog may be the subject of pawn, in which case the pawnee is bound to use ordinary diligence in keeping it. He will not be responsible if the pawn be stolen from him, unless it can be shown that he has been guilty of negligence. Neither will he be responsible if the pawn be destroyed by fire, unless it can be shown that the fire took place through his negligence or default. He has not, in general, any right to use the thing pledged, unless such use were beneficial to it, when the consent of the pawnor to the use of the article pledged would be implied. And where the pawn is of such a nature as the pawnee is at any charge about the thing pawned to maintain it, as a horse, cow, &c., then he may use the horse in a reasonable manner, or milk the cow, in recompense for the keeping (*w*). In like manner, it would seem that he may use a dog for the purpose of sporting, &c.

(*w*) *Coggs* v. *Bernard*, 2 Ld. Raym. 916, 917; Smith's Leading Cases.

CHAPTER VIII.

DOG STEALING, AND SALES OF STOLEN DOGS.

THE offence of dog stealing is provided for by the 24 & 25 Vict. c. 96, which renders it punishable, on summary conviction for the first offence, by imprisonment, with or without hard labour, for a term not exceeding *six calendar months*, or a fine not exceeding *twenty pounds* beyond the value of the dog. A subsequent offence is an indictable misdemeanour, punishable by imprisonment for a term not exceeding *eighteen calendar months*, with or without hard labour (*x*).

Persons found in possession of dogs, or the skins of dogs, knowing them to have

(*x*) The Act, s. 18. On appearing before the court, and before the charge is gone into, the accused may claim to be tried by a jury. 42 & 43 Vict. c. 49, s. 17, *post*, p. 124. As the offence of dog stealing does not amount to larceny, there cannot be a conviction for obtaining a dog by false pretences under this section. *Reg.* v. *Robinson*, 28 L. J. M. C. 58.

been stolen, are liable, for a first offence, to a fine not exceeding the sum of *twenty pounds*, and for a second offence, which is also an indictable misdemeanour, to imprisonment, with or without hard labour, for a term not exceeding *eighteen calendar months* (*y*). An order may be made for the restitution of a stolen dog, or the skin thereof, to the owner (*z*). All sums of money forfeited on conviction are to be paid to the aggrieved party (*a*); and the offender, in default of paying the same, is liable to be imprisoned, with or without hard labour (*b*), as specified in the 42 & 43 Vict. c. 49, s. 5 (*c*).

If, in advertising a reward for the return of a lost or stolen dog, words are used purporting that no question will be asked, or that a reward will be paid, without seizing or making inquiry after person producing the animal, the advertiser, and the printer and the publisher of the advertisement, will forfeit the sum of 50*l.* to any person who will

(*y*) 24 & 25 Vict. c. 96, s. 19. See *post*, p. 124.

(*z*) *Idem*, s. 22.

(*a*) *Idem*, s. 106.

(*b*) *Idem*, s. 107.

(*c*) See *post*, p. 123.

sue for the same by action of debt, to be recovered with full costs of suit (*e*).

Apprehension without warrant.

Any person found committing an offence may be immediately apprehended without a warrant by any person; and if any credible witness prove on oath, before a justice of the peace, a reasonable cause to suspect that any person has in his possession or on his premises any stolen dog, or the skin thereof, the justice may grant a warrant to search for the same. And any person to whom it may be offered for sale or delivery, having reasonable cause to suspect that it has been stolen, is authorized, and, if in his power, required, to apprehend the party offering the same (*f*).

Justice may grant search warrant.

Person to whom stolen property is offered may apprehend party offering it.

Justice may discharge offender in case of first conviction.

In the case of a first conviction the justice may discharge the offender upon his making satisfaction to the aggrieved party for damages and costs (*g*).

Summary conviction a bar to other proceedings.

A summary conviction will be a bar to any other proceeding for the same cause (*h*).

Appeal.

Where the sum adjudged to be paid on

(*e*) 24 & 25 Vict. c. 96, s. 102.

(*f*) *Idem*, s. 103.

(*g*) *Idem*, s. 108.

(*h*) *Idem*, s. 109.

any summary conviction exceeds *five pounds*, or the imprisonment exceeds *one month*, or the conviction takes place before one justice only, any person who thinks himself aggrieved may appeal to the next court of general or quarter sessions for the county or place wherein the cause of complaint arose (*i*).

Sales of stolen dogs.

It remains for us to consider the effect of the sale of a dog by a thief or person having wrongful possession thereof. As a general rule, a man who has no property in goods cannot, by making a sale, transfer the property therein to another, and an action is maintainable to recover the value of the goods which have been stolen from the plaintiff, and which the defendant had innocently purchased, although no steps have been taken to bring the thief to justice. But the owner is not allowed (on grounds of public policy) to bring an action for them against the thief, or against him and others jointly, until after conviction (*j*).

By sect. 100 of the 24 & 25 Vict. c. 96, above mentioned, if any person guilty of any

(*i*) *Idem*, s. 110. See also 42 & 43 Vict. c. 49, ss. 31, 32.

(*j*) *White* v. *Spettigue*, 13 M. & W. 606; *Lee* v. *Bayes*, 18 C. B. 602.

such misdemeanour as is mentioned in the act, which includes the misdemeanour of dog stealing, &c., shall be indicted by or on behalf of the owner of the property, and convicted thereof, in such case the property shall be restored to the owner or his representative, and the court before whom any such person is tried for any such misdemeanour has power to award, from time to time, writs of restitution for the property, or to order the restitution thereof in a summary manner.

Market overt.

A thief or person in wrongful possession of goods may make a sale of them in market overt (*k*) to a *bonâ fide* purchaser who does not know that they are not his property, and such sale will be valid as between himself and the purchaser; but as regards the rightful owner, the sale, though it will not be

(*k*) Market overt in the country is only held on the special days provided for particular towns by charter or prescription; but in London every day, except Sunday, is market day. The market place, or spot of ground set apart by custom for the sale of particular goods, is also, in the country, the only market overt; but in London, every shop in which goods are exposed publicly for sale is market overt for such things only as the owner professes to trade in. Sale in market overt must be held in a place that is open, so that every one who passes may see it, and that it is proper for the sale of such goods; it should take place between sunrise and sunset.

absolutely void, will, nevertheless, be voidable (*l*).

Accordingly, it has been recently held by the Court of Appeal, that if a contract for the sale of goods has been obtained by false pretences, and the owner of the goods has prosecuted the misdemeanant to conviction, the property in the goods re-vests in the owner at *the time of conviction*, and he can then recover them from any person in whose possession they are, even though that person purchased them *bonâ fide* in market overt, or without notice of the fraud. As the restoration of the property to the original owner is to date from the conviction, it follows that it is to be restored only as against the person in whose possession the goods are at the time of conviction, and that a *bonâ fide* purchaser is not responsible for any dealings with the goods prior to the conviction (*m*). So where goods having been stolen were sold in market overt to the defendant, who sold them again, and after that the thief was

(*l*) *Horwood* v. *Smith*, 2 T. R. 750; *Vilmont* v. *Bentley*, 56 L. T. R. (N. S.) 318, overruling *Moyce* v. *Newington*, 39 L. T. R. (N. S.) 535, and distinguishing *Horwood* v. *Smith*, *supra*.

(*m*) *Vilmont* v. *Bentley*, *supra*.

convicted, the question arose, whether the original owner of the goods could claim them against the defendant, who had bought the goods in market overt and sold them before conviction, and the court held that he could not (*n*).

The court has power to order restitution not only of the goods themselves, but also of the proceeds of their sale in the hands of the thief, and if they are in the hands of an agent who holds them from him, that will be sufficient (*o*).

(*n*) *Horwood* v. *Smith*, 2 T. R. 750, *supra*.

(*o*) *Reg.* v. *Justices of the Central Criminal Court*, 56 L. T. R. (N. S.) 352, and cases there cited.

CHAPTER IX.

EXCISE DUTY ON DOGS.

Duty on dogs at different periods.

In this chapter we shall say a few words concerning the duty payable on dogs.

Duty was first imposed in 1796, and levied as an assessed tax. It varied in amount from time to time, and different rates were charged, according to the kind of dog; but in 1853, on the revision of the assessed taxes, the duty was fixed at *twelve shillings*, and this was the amount at which it stood when last collected as an assessed tax. In 1867 the duty was fixed at *five shillings* for every description of dog, and so continued until 1878, when it was fixed at *seven shillings and sixpence*, which amount is now payable annually in respect of every dog by the person who keeps the animal, subject to certain exceptions which we shall presently notice (*p*). The duty is now regulated by the 30 & 31 Vict.

(*p*) 42 & 43 Vict. c. 21, s. 26.

c. 5 (called the principal Act); the 41 & 42 Vict. c. 15, ss. 17—23; and the 42 & 43 Vict. c. 21, s. 26.

Who to be deemed the keeper of a dog.

The person in whose custody, charge, or possession, or in whose house or premises any dog is found or seen, will be deemed the person who keeps the animal, unless the contrary be proved. And the owner or master of the hounds will be deemed the person keeping the same (*q*).

In certain cases to be noticed presently, where exemptions are allowed, a declaration must be made. The form of this declaration may be obtained from the supervisor or excise officer of the district in which the person claiming exemption resides. It must be filled up and signed by him, and then returned to the officer. If the claim is allowed, the required certificate will be forwarded in due course.

Penalty for delivering false declaration.

A person delivering a false declaration is liable to a penalty of *twenty pounds;* and a person to whom a certificate of exemption may have been issued not producing the same within a reasonable time after the

(*q*) See *ante*, p. 7.

request of any officer of excise or police is liable to a penalty of *five pounds* (*s*).

The taking out of a licence after the offence is discovered will not exempt him from liability. Therefore, where an excise officer called at the house of the defendant at about 12.10 p.m., and there saw a dog kept by him, and no licence for the dog was then in force, but on the same day at 1.10 p.m. the defendant took out a licence authorizing him to keep one dog "*from the date hereof*" until December 31st, 1877, it was held that the subsequent licence operated only from the time when it was granted, and did not relate back to the earliest moment of that day, so as to justify the violation of the act before the licence existed (*t*). The defendant is not a competent witness in such cases (*u*).

Penalty for keeping dog without a licence.

Any person keeping a dog without a licence, or keeping a greater number of dogs than he is licensed to keep, is liable to forfeit the sum of *five pounds* (*v*).

(*s*) 30 & 31 Vict. c. 5, s. 8, sub-s. 4.

(*t*) *Campbell* v. *Strangways*, L. R. 3 C. P. Div. 105 ; 47 L. J. M. C. 6.

(*u*) *Regina* v. *Sullivan*, 19 S. J. 235.

(*v*) 30 & 31 Vict. c. 5, s. 7.

Licences. On payment of the duty to the proper authority, a licence will be issued in due form. It is not transferable. This licence, which may be obtained at any money-order office, will commence from the moment when granted, and terminate on the 31st of December following (*w*). The Commissioners of Inland Revenue are to cause to be placed upon or near to the door of every church in Great Britain, a printed or written notice, stating from whom licences can be obtained by persons residing in the parish or place in which such church is situated, and every such notice must be kept affixed upon or near the door of the church for such time as the Commissioners may direct; but no proceeding of any kind, nor any act done by any person in pursuance of the act, will be deemed invalid or unlawful by reason of the notice not having been placed or affixed as aforesaid (*x*).

A register of all licences, wherever granted, is kept by the Excise officer of

(*w*) 30 & 31 Vict. c. 5, s. 5; *Campbell* v. *Strangways*, L. R. 3 C. P. Div. 105; 47 L. J. M. C. 6.

(*x*) 30 & 31 Vict. c. 5, s. 7.

the station in which the person who has taken out a licence resides (*y*).

Exemptions from duty are allowed in the following cases:—

(1) Dogs under the age of six months (*z*). (2) Hounds under the age of twelve months, which have never been entered or used with any pack of hounds (*a*). Proof of age lies on the owner (*b*). (3) Dogs used solely for the guidance of blind persons (*c*). (4) Farmers and shepherds will, upon filling up and signing the form of declaration as prescribed by the Commissioners of Inland Revenue, be entitled to receive a certificate of exemption in respect of a dog or dogs, not exceeding two in number, solely used for tending sheep or cattle, or in the exercise of the calling of a shepherd (*d*).

Exemptions from duty.

Declaration by farmers and shepherds.

Where an occupier of a sheep-farm owns more than 400 sheep which feed on common or unenclosed land, so that more than *two* dogs are required for the purpose of tending

(*y*) 30 & 31 Vict. c. 5, s. 6.
(*z*) *Idem*, s. 10.
(*a*) *Idem*, s. 20.
(*b*) *Idem*, s. 19.
(*c*) *Idem*, s. 21.
(*d*) *Idem*, s. 22.

sheep, and such facts are stated in the declaration delivered by him, he will be entitled to a certificate of exemption in respect of a third dog kept by him solely for that purpose; and if the number of his sheep amounts to 1,000, then in respect of a fourth dog so kept, and in respect of an additional dog so kept for every full number of 500 sheep so owned by him over and above the number of 1,000. But he will not be entitled to exemption in respect of more than *eight* dogs on the farm (*e*); though in certain cases, where the number of sheep is very large, the Board of Inland Revenue allow exemption in respect of more than eight dogs.

Dog used for taking rats, &c. not exempt.

If a dog is used by a farmer or shepherd in taking rats, rabbits, or game, or for any other purpose than that of solely tending sheep or cattle, he will not be entitled to exemption.

Butchers and drovers not exempt.

Butchers and drovers who occupy land for their trade are not entitled to exemption. So, a butcher, or drover, or auctioneer who holds grass parks, and keeps a dog for driving cattle to and from such parks, cannot claim exemption; neither can a cattle salesman or dairyman.

(*e*) 30 & 31 Vict. c. 5, s. 22, sub-s. 2.

But a farmer who uses his dog occasionally to drive cattle or sheep to market may claim.

Further exemptions.

The authorities do not require licences to be taken out in the following cases:—Tenant cottars having unenclosed land using a dog to guard their own cattle or sheep, or to scare the cattle or sheep of other persons straying on the ground. Tenant cottars holding unenclosed land in common may claim exemption for a herd's dog, but in such case individual cottars are not entitled to claim. The steward or manager of a proprietor or farmer not residing on the farm, and also the legal representatives, trustees, &c., of a farmer, may claim exemption.

Penalty for not producing licence.

Any person not producing his licence to any officer of excise or police-constable within a reasonable time after request is liable to forfeit the sum of *five pounds* (*f*), and in default of payment may be imprisoned for *one calendar month* (*g*).

Recovery of penalties.

Penalties may be sued for, and recovered or enforced upon the information of a police-

(*f*) *Idem*, s. 9.
(*g*) 42 & 43 Vict. c. 49, s. 5.

constable (*h*), in a court of summary jurisdiction, which has power to award costs and to mitigate the penalty to such an amount as it may in its discretion think fit in the case of a first offence (*i*).

Information may also be laid by an excise officer (*j*).

An appeal lies to the court of quarter sessions.

(*h*) 41 & 42 Vict. c. 15, s. 23.

(*i*) *Idem.* But see 42 & 43 Vict. c. 49, s. 4, *post*, p. 123. Formerly the court had no power to mitigate penalties, though it might have recommended a reduction thereof to the Commissioners of Inland Revenue, who may mitigate or wholly remit them. See 7 & 8 Geo. 4, c. 53, s. 78.

(*j*) 4 & 5 Will. 4, c. 5, s. 19.

CHAPTER X.

PROCEEDINGS IN COURTS OF SUMMARY JURISDICTION.

Proceedings in courts of summary jurisdiction.

As many of the cases which we have had occasion to notice in the course of the work are cognizable in the courts of summary jurisdiction, a few observations on the constitution, powers, and procedure of such courts may be found useful. Cases are there heard and determined by one or more justice or justices of the peace, or by a stipendiary magistrate, the latter being empowered to do alone whatever is authorized to be done by one or more of the former (*p*). It must, therefore, be understood, that when we speak of a justice or justices of the peace the expression equally applies to a stipendiary magistrate sitting alone. The mode of instituting proceedings is frequently indicated in the statute under which each particular

(*p*) 2 & 3 Vict. c. 71, s. 14; 11 & 12 Vict. c. 43, s. 33.

offence is created, and will be commonly regulated by the Summary Jurisdiction Acts, 1848 and 1879. Offences are punishable by fine or imprisonment, to the amount or extent named in the several acts, subject, however, to the provisions of the 42 & 43 Vict. c. 49, which gives the court increased discretionary powers in awarding punishments. It may now, notwithstanding anything contained to the contrary in any past or future act, impose imprisonment or a fine; and, in the case of imprisonment, may impose it *without* hard labour, and reduce the prescribed period thereof, or do either of such acts; and, in the case of a fine, if it be imposed in respect of a first offence, may reduce the prescribed amount thereof. And where the court has authority under any past or future act to impose imprisonment, and no authority to impose a fine, it may, if it think the justice of the case will be better met by a fine than imprisonment, impose a fine not exceeding *twenty-five pounds*, and not being of such an amount as will subject the offender, in default of payment, to any greater term of imprisonment than that to which he is liable under the act authorizing the imprison-

42 & 43 Vict. c. 49 (Summary Jurisdiction Act, 1879).

ment (*q*). The period of imprisonment in respect of the non-payment of any sum of money adjudged to be paid by a conviction, or in respect of the default of a sufficient distress to satisfy any such sum, will be such period as in the opinion of the court will satisfy the justice of the case, but must not exceed the maximum fixed by the following scale, that is to say—

Scale of punishments.

Where the amount of the sum or sums of money adjudged to be paid by a conviction, as ascertained by the conviction,	The said period shall not exceed
Does not exceed 10*s*. - - - -	Seven days.
Exceeds 10*s*., but does not exceed 1*l*. -	Fourteen days.
Exceeds 1*l*., but does not exceed 5*l*. -	One month.
Exceeds 5*l*. but does not exceed 20*l*. -	Two months.
Exceeds 20*l*. - - - - -	Three months.

And such imprisonment must be *without* hard labour, except where hard labour is authorized by the act on which the conviction is founded, in which case the imprisonment may, if the court thinks the justice of the case requires it, be with hard labour, so that the term of hard labour awarded do not exceed the term authorized by the act (*r*).

(*q*) 42 & 43 Vict. c. 49, s. 4.

(*r*) *Idem*, s. 5. Where fine or imprisonment with hard labour can be inflicted, imprisonment with hard labour can be inflicted in default of payment of the fine. *Reg.* v. *Justices of Tynemouth*, 55 L. J. M. C. 181.

Civil debts recoverable in courts of summary jurisdiction.

A sum of money claimed to be due on complaint, and not on information, is deemed to be a civil debt, and recoverable summarily as such, together with any costs ordered to be paid by the complainant or defendant (*s*).

The court may allow time for payment of any sum adjudged to be paid, and direct it to be paid by instalments, or that the person liable give security for the same.

Right to be tried by jury.

Where a person is charged with an offence for which he is liable to be imprisoned for a term exceeding three months (*t*) (and which is not an assault), he may, on appearing before the court, and before the charge is gone into, but not afterwards, claim to be tried by a jury, when the case must be dealt with in all respects as if the accused were charged with an indictable offence (*u*).

Case heard in occasional

Where a case is heard and determined in an "occasional court-house" (*v*), the period

(*s*) 42 & 43 Vict. c. 49, s. 6.

(*t*) The word "*month*" means calendar month, unless words be added showing lunar to be intended. 13 & 14 Vict. c. 21, s. 4.

(*u*) 42 & 43 Vict. c. 49, s. 17.

(*v*) An "occasional court-house" means a police station, or other place for the time being appointed to be used as an occasional court-house. The Act, s. 20, sub-s. 4, s. 50.

of imprisonment must not exceed *fourteen* days, nor must the sum adjudged to be paid exceed *twenty shillings;* and a justice of the peace sitting alone in a petty sessional court-house (*w*) has no power to impose a greater term of imprisonment, or adjudge any larger sum to be paid, than is above mentioned (*x*). court-house. Justice sitting alone.

Where a person who is aggrieved by a conviction or order desires to appeal therefrom, the act on which the conviction is founded should be examined, with a view to ascertain whether it gives a right of appeal; and where this is found to be the case, the conditions and regulations thereof should be carefully studied, as well as the clauses relative thereto contained in the Act now under consideration (*y*). Appeal.

We have thus endeavoured to give a brief outline of the chief provisions of the Sum-

(*w*) A "petty sessional court-house" means a court-house or other place at which justices are accustomed to assemble for holding special or petty sessions, or which is for the time being appointed as a substitute for such court-house or place; and where the justices are accustomed to assemble for either special or petty sessions at more than one court-house or place in a petty sessional division, it means any such court-house or place. *Idem*, s. 20, sub-s. 3.

(*x*) *Idem*, sub-s. 7.

(*y*) *Idem*, ss. 31, 32.

mary Jurisdiction Acts, which it is essential to bear in mind for the purposes of this work. Attention has been already called to various other points in connection with these acts, both in the text and foot-notes.

APPENDIX.

STATUTES (IN FULL).

28 & 29 VICT. c. 60 (*Injuries to Cattle and Sheep*).

34 & 35 VICT. c. 56 (*Dogs Act*, 1871).

39 & 40 VICT. c. 77 (*Vivisection*).

---

28 & 29 VICT. c. 60.

An Act to render Owners of Dogs in England and Wales liable for Injuries to Cattle and Sheep.

[29th June, 1865.]

WHEREAS it is expedient to amend the law as to the liability of the owners of dogs for injuries done to cattle and sheep by such dogs: Be it therefore enacted by the Queen's most excellent Majesty, by and with the advice and consent of the Lords Spiritual and Temporal, and Commons, in this present Parliament assembled, and by the authority of the same, as follows:

Owner of dog to be liable in damages for any injury committed by his dog.

1. The owner of every dog shall be liable in damages for injury done to any cattle or sheep by his dog; and it shall not be necessary for the party seeking such damages to show a previous mischievous propensity in such dog, or the owner's knowledge of such previous propensity, or that the injury was attributable to neglect on the part of such owner. Such damages shall be recoverable in any court of competent jurisdiction by the

Recovery of damages.

owner of such cattle or sheep killed or injured. Where the amount of the damages claimed shall not exceed five pounds, the same shall be recoverable in a summary way before any justice or justices sitting in petty sessions under the provisions of the Act eleven and twelve Victoria, chapter forty-three.

Who shall be deemed the owner of the dog.

2. The occupier of any house or premises where any dog was kept or permitted to live or remain at the time of such injury shall be deemed to be the owner of such dog, and shall be liable as such, unless the said occupier can prove that he was not the owner of such dog at the time the injury complained of was committed, and that such dog was kept or permitted to live or remain in the said house or premises without his sanction or knowledge: Provided always, that where there are more occupiers than one in any house or premises let in separate apartments, or lodgings, or otherwise, the occupier of that particular part of the premises in which such dog shall have been kept or permitted to live or remain at the time of such injury shall be deemed to be the owner of such dog.

Extent of act.

3. This Act shall extend to England and Wales only.

34 & 35 VICT. c. 56.

An Act to provide further Protection against Dogs.

[24th July, 1871.]

WHEREAS it is expedient that further protection should be provided against dogs:

Be it enacted by the Queen's most excellent Majesty, by and with the advice and consent of the Lords Spiritual and Temporal, and Commons, in this present Parliament assembled, and by the authority of the same, as follows:

Stray dogs may be detained and sold or destroyed.

1. From and after the passing of this act any police officer or constable may take possession of any dog that he has reason to suppose to be savage or dangerous straying on any highway, and not under the control of any person, and may detain such dog until the owner has claimed the same, and paid all expenses incurred by reason of such detention.

Where the owner of any dog taken possession of by any constable is known, a letter, stating the fact of such dog having been taken possession of, shall be sent by post or otherwise to the owner at his usual or last known place of abode.

When any dog taken in pursuance of this act has been detained for three clear days where the owner is not known as aforesaid, or for five clear days where he is so known, without the owner claiming the same, and paying all expenses incurred by its detention, the chief officer of police of the district in which such dog was found may cause such dog to be sold or destroyed.

Any moneys arising from the sale of any dogs in pursuance of this section shall be paid to the account of the local rate, and be applied to the purposes to which that rate is applicable.

All dogs detained under this section shall be properly fed and maintained at the expense of the local rate.

Dangerous dogs may be destroyed.

2. Any court of summary jurisdiction may take cognizance of a complaint that a dog is dangerous, and not kept under proper control, and if it appears to the court having cognizance of such complaint that such dog is dangerous, the court may make an order in a summary way directing the dog to be kept by the owner under proper control or destroyed, and any person failing to comply with such order shall be liable to a penalty not exceeding twenty shillings for every day during which he fails to comply with such order.

Restriction upon dogs being at large if danger from mad dogs is apprehended.

3. The local authority may, if a mad dog, or a dog suspected of being mad, is found within their jurisdiction, make, and when made vary or revoke, an order placing such restrictions as they think expedient on all dogs not being under the control of any person during such period as may be prescribed in such order throughout the whole of their jurisdiction, or such part thereof as may be prescribed in such order.

Any person who acts in contravention of any order made in pursuance of this section shall be liable to a penalty not exceeding twenty shillings.

Due notice of such order shall be published, at the expense of the local rate.

The provisions in this act contained as to the detention and sale or destruction of dogs found straying on the highway shall apply to dogs found at large in contravention of any order made in pursuance of this section.

4. In England and Ireland any penalty under this act may be recovered in manner provided by the Summary Jurisdiction Acts; and in Scotland all such penalties shall be prosecuted and recovered before a court of summary jurisdiction, under the provisions of the Summary Jurisdiction Act, 1864. Penalty how to be recovered.

5. In this act,— Definition of terms.

For the purposes of this act and in reference to the districts mentioned in the first column of the schedule annexed hereto, "local authority" and "local rate" mean the bodies of persons mentioned in the second column and the rate mentioned in the third column of the said schedule:

"Summary Jurisdiction Acts" mean as follows:

As to England, the act of the session of the eleventh and twelfth years of the reign of her present Majesty, chapter forty-three, intituled "An Act to facilitate the performance of the duties of justices of the peace out of sessions within England and Wales with respect to summary convictions and orders," and any acts amending the same;

As to Scotland, "The Summary Procedure Act, 1864:"

As to Ireland, within the police district of Dublin metropolis, the acts regulating the powers and duties of justices of the peace for such district, or of the police of such district; and elsewhere in Ireland, "The Petty Sessions (Ireland) Act, 1851," and any act amending the same:

"Court of summary jurisdiction" means, in England and Ireland, any two justices of the peace or any metropolitan stipendiary or other magistrate empowered by law to do alone, or with others, any act authorized to be done by more than one justice of the peace; and in Scotland, any justice or justices of the peace, sheriff or sheriff substitute, police or other magistrate, or officer, by whatever name called, to whom jurisdiction is given, or proceedings before whom may be regulated, by the Summary Jurisdiction Acts or any acts therein referred to:

The expression "police district" means,—

In England,—

1. The city of London and the liberties thereof;
2. The metropolitan police district;
3. Any county, riding, part, division, or liberty of a county, borough, city, town, place, or union, or combination of places maintaining a separate police force; and all the police under one chief constable shall be deemed to constitute one force for the purposes of this definition:

In Scotland,—

Any area maintaining a separate police force; and all the police under one chief constable shall be deemed to constitute one force for the purposes of this definition:

In Ireland,—

1. The police district of Dublin metropolis;
2. Any district, whether city, town, or country, over which is appointed a sub-inspector of the Royal Irish Constabulary:

The expression "chief officer of police" means,—

In England,—

1. In the city of London and the liberties thereof, the Commissioner of City Police;
2. In the metropolitan police district, the Commissioner of Police of the Metropolis;

3. Elsewhere, the chief constable, or head constable, or other officer, by whatever name called, having the chief command of the police in the police district in reference to which such expression occurs:

In Scotland,—

The chief constable, superintendent of police, or other officer, by whatever named called, having the chief command of the police in the police district in reference to which such expression occurs:

In Ireland,—

1. In the police district of Dublin metropolis, either of the commissioners of police for the said district;
2. In any other police district, the sub-inspector of the Royal Irish Constabulary:

Any act or thing by this act authorized to be done by the chief officer of police may be done by any person authorized by him in that behalf:

"Highway" includes any street or any place of public resort.

6. This act shall not affect the powers contained in the eighteenth section of the Metropolitan Streets Act, 1867, or in any local or other act of parliament for the same or like purposes; and in places where any such local or other act is in force, proceedings may be taken under such local or other act, or under this act, as may be deemed expedient. Saving of local acts and of sect. 18 of 30 & 31 Vict. c. 134.

7. This act may be cited as "The Dogs Act, 1871." Short title.

SCHEDULE.

ENGLAND.

District of Local Authority.	Description of Local Authority.	Local Rate.
City of London and the liberties thereof.	The Commissioner of City Police.	The rate applicable to maintenance of police.
Metropolitan police district.	The Commissioner of Police for the Metropolis.	Do.
Any borough subject to the act of the session of the fifth and sixth years of the reign of King William the Fourth, chapter seventy-six, intituled "An Act to provide for the regulation of Municipal Corporations in England and Wales."	The mayor, aldermen, and burgesses acting by the town council.	Do.
Any district (no portion of which is included in a borough) for the time being subject to the jurisdiction of any commissioners, trustees, or other persons intrusted by any local act, not being a turnpike act or a highway act, with powers of improving, cleansing, or paving any part of such district.	The commissioners, trustees, or other body of persons so intrusted.	Do.
Places in England not included in the above description.	The justices in petty sessions assembled at their usual place of meeting.	Do.

SCOTLAND.

District of Local Authority.	Description of Local Authority.	Local Rate.
Burghs, Royal or Parliamentary, not subject to the separate jurisdiction of police commissioners or trustees.	The town council .	The revenue of the burgh or any rate applicable to sewers leviable by the town council.
Burghs and places where police commissioners or trustees exercise the functions of police commissioners or trustees under any general or local act.	The police commissioners.	Any rate leviable by the commissioners or trustees, or any fund belonging to them.
Any place in Scotland not included in the above description.	The sheriff or sheriff substitute.	The rate applicable to the maintenance of police.

IRELAND.

District of Local Authority.	Description of Local Authority.	Local Rate.
The city of Dublin	The Right Honourable the Lord Mayor, aldermen, and burgesses acting by the town council.	The borough rate or borough fund.
Towns corporate, with the exception of Dublin.	The mayor, aldermen, and burgesses acting by the town council.	
Towns having commissioners appointed by virtue of an act made in the ninth year of the reign of George the Fourth, intituled "An Act to make provision for the Lighting, Cleansing, and Watching of Cities, and Towns Corporate and Market Towns in Ireland in certain cases."	The commissioners	Any rate levied by the commissioners.
Towns having town commissioners under the Towns Improvement (Ireland) Act, 1854 (17 & 18 Vict. c. 103), or under any other local act.	The town commissioners.	
Townships having commissioners under local acts.	The township commissioners.	
Places in Ireland not included in the foregoing descriptions.	The justices of the peace for the petty sessions of the district in which such place is situate.	The grand jury cess.

39 & 40 VICT. C. 77.

An Act to amend the Law relating to Cruelty to Animals. [15th August, 1876.]

WHEREAS it is expedient to amend the law relating to cruelty to animals by extending it to the cases of animals which for medical, physiological, or other scientific purposes are subjected when alive to experiments calculated to inflict pain:

Be it enacted by the Queen's most excellent Majesty, by and with the advice and consent of the Lords Spiritual and Temporal, and Commons, in this present Parliament assembled, and by the authority of the same, as follows:

1. This act may be cited for all purposes as "The Cruelty to Animals Act, 1876." Short title.

2. A person shall not perform on a living animal any experiment calculated to give pain, except subject to the restrictions imposed by this act. Any person performing or taking part in performing any experiment calculated to give pain, in contravention of this act, shall be guilty of an offence against this act, and shall, if it be the first offence, be liable to a penalty not exceeding fifty pounds, and if it be the second or any subsequent offence, be liable, at the discretion of the court by which he is tried, to a penalty not exceeding one hundred pounds, or to imprisonment for a period not exceeding three months. Prohibition of painful experiments on animals.

General restrictions as to performance of painful experiments on animals.

3. The following restrictions are imposed by this act with respect to the performance on any living animal of an experiment calculated to give pain; that is to say,

(1.) The experiment must be performed with a view to the advancement by new discovery of physiological knowledge or of knowledge which will be useful for saving or prolonging life, or alleviating suffering; and

(2.) The experiment must be performed by a person holding such license from one of Her Majesty's Principal Secretaries of State, in this act referred to as the Secretary of State, as is in this act mentioned, and in the case of a person holding such conditional license as is hereinafter mentioned, or of experiments performed for the purpose of instruction in a registered place; and,

(3.) The animal must, during the whole of the experiment, be under the influence of some anæsthetic of sufficient power to prevent the animal feeling pain; and

(4.) The animal must, if the pain is likely to continue after the effect of the anæsthetic has ceased, or if any serious injury has been inflicted on the animal, be killed before it recovers from the influence of the anæsthetic which has been administered; and

(5.) The experiment shall not be performed as an illustration of lectures in medical schools, hospitals, colleges, or elsewhere; and

(6.) The experiment shall not be performed for the purpose of attaining manual skill.

Provided as follows; that is to say,

(1.) Experiments may be performed under the foregoing provisions as to the use of anæsthetics by a person giving illustrations of lectures in medical schools, hospitals, or colleges, or elsewhere, on such certificate being given as in this act mentioned, that the proposed experiments are absolutely necessary for the due instruction of the persons to whom such lectures are given with a view to their acquiring physiological knowledge or knowledge which will be useful to them for saving or prolonging life or alleviating suffering; and

(2.) Experiments may be performed without anæsthetics on such certificate being given as in this act mentioned that insensibility cannot be produced without necessarily frustrating the object of such experiments; and

(3.) Experiments may be performed without the person who performed such experiments being under an obligation to cause the animal on which any such experiment is performed to be killed before it recovers from the influence of the anæsthetic on such certificate being given as in this act mentioned, that the so killing the animal would necessarily frustrate the object of the experiment, and provided that the animal be killed as soon as such object has been attained; and

(4.) Experiments may be performed not directly for the advancement by new discovery of physiological knowledge, or of knowledge which will be useful for saving or prolonging life or alleviating suffering, but for the purpose of testing a particular former discovery alleged to have been made for the advancement of such knowledge as last aforesaid, on such certificate being given as is in this act mentioned that such testing is absolutely necessary for the effectual advancement of such knowledge.

Use of urari as an anæsthetic prohibited.

4. The substance known as urari or curare shall not for the purposes of this act be deemed to be an anæsthetic.

Special restrictions on painful experiments on dogs, cats, &c.

5. Notwithstanding anything in this act contained, an experiment calculated to give pain shall not be performed without anæsthetics on a dog or cat, except on such certificate being given as in this act mentioned, stating, in addition to the statements hereinbefore required to be made in such certificate, that for reasons specified in the certificate the object of the experiment will be necessarily frustrated unless it is performed on an animal similar in constitution and habits to a cat or dog, and no other animal is available for such experiment; and an experiment calculated to give pain shall not be performed on any horse, ass, or mule except on such certificate being given as in this act mentioned that the object of the experiment will be necessarily frustrated unless it is performed on a horse, ass, or mule, and that no other animal is available for such experiment.

Absolute prohibition of public exhibition of painful experiments.

6. Any exhibition to the general public, whether admitted on payment of money or gratuitously, of experiments on living animals calculated to give pain shall be illegal.

Any person performing or aiding in performing such experiments shall be deemed to be guilty of an offence against this act, and shall, if it be the first offence, be liable to a penalty not exceeding fifty pounds, and if it be the second or any subsequent offence, be liable, at the discretion of the court by which he is tried, to a penalty not exceeding one hundred pounds or to imprisonment for a period not exceeding three months.

And any person publishing any notice of any such intended exhibition by advertisement in a newspaper, placard, or otherwise shall be liable to a penalty not exceeding one pound.

A person punished for an offence under this section shall not for the same offence be punishable under any other section of this act.

Administration of Law.

Registry of place for performance of experiments.

7. The Secretary of State may insert, as a condition of granting any license, a provision in such license that the place in which any experiment is to be performed by the licensee is to be registered in such manner as the Secretary of State may from time to time by any general or special order direct; provided that every place for the performance of experiments for the purpose of instruction under this act shall be approved by the Secretary of State, and shall be registered in such manner as he may from time to time by any general or special order direct.

License by Secretary of State.

8. The Secretary of State may license any person whom he may think qualified to hold a license to perform experiments under this act. A license granted by him may be for such time as he may think fit, and may be revoked by him on his being satisfied that such license ought to be revoked. There may be annexed to such license any conditions which the Secretary of State may think expedient for the purpose of better carrying into effect the objects of this act, but not inconsistent with the provisions thereof.

Reports to Secretary of State.

9. The Secretary of State may direct any person performing experiments under this act from time to time to make such reports to him of the result of such experiments, in such form and with such details as he may require.

Inspection by Secretary of State.

10. The Secretary of State shall cause all registered places to be from time to time visited by inspectors for the purpose of securing a compliance with the provisions of this act, and the Secretary of State may, with the assent of the Treasury as to number, appoint any special inspectors, or may from time to time assign the duties of any such inspectors to such officers in the employment of the government, who may be willing to accept the same, as he may think fit, either permanently or temporarily.

Certificate of scientific bodies for exceptions to general regulations.

11. Any application for a license under this act and a certificate given as in this act mentioned must be signed by one or more of the following persons; that is to say,

The President of the Royal Society;

The President of the Royal Society of Edinburgh;

The President of Royal Irish Academy;

The Presidents of the Royal Colleges of Surgeons in London, Edinburgh, or Dublin;

The Presidents of the Royal Colleges of Physicians in London, Edinburgh, or Dublin;

The President of the General Medical Council;

The President of the Faculty of Physicians and Surgeons of Glasgow;

The President of the Royal College of Veterinary Surgeons, or the President of the Royal Veterinary College, London, but in the case only of an experiment to be performed under anæsthetics with a view to the advancement by new discovery of veterinary science;

and also (unless the applicant be a professor of physiology, medicine, anatomy, medical jurisprudence, materia medica, or surgery in a university in Great Britain or Ireland, or in University College, London, or in a college in Great Britain or Ireland, incorporated by royal charter) by a professor of physiology, medicine, anatomy, medical jurisprudence, materia medica, or surgery in a university in Great Britain or Ireland, or in University College, London, or in a college in Great Britain or Ireland, incorporated by royal charter.

Provided that where any person applying for a certificate under this act is himself one of the persons authorized to sign such certificate, the signature of some other of such persons shall be substituted for the signature of the applicant.

A certificate under this section may be given for such time or for such series of experiments as the person or persons signing the certificate may think expedient.

A copy of any certificate under this section shall be forwarded by the applicant to the Secretary of State, but shall not be available until one week after a copy has been so forwarded.

The Secretary of State may at any time disallow or suspend any certificate given under this section.

Power of judge to grant license for experiment when necessary in criminal case.

12. The powers conferred by this act of granting a license or giving a certificate for the performance of experiments on living animals may be exercised by an order in writing under the hand of any judge of the High Court of Justice in England, of the High Court of Session in Scotland, or of any of the superior courts in Ireland, including any court to which the jurisdiction of such last-mentioned courts may be transferred, in a case where such judge is satisfied that it is essential for the purposes of justice in a criminal case to make any such experiment.

Legal Proceedings.

Entry on warrant by justice.

13. A justice of the peace, on information on oath that there is reasonable ground to believe that experiments in contravention of this act are being performed by an unlicensed person in any place not registered under this act may issue his warrant authorizing any officer or constable of police to enter and search such place, and to take the names and addresses of the persons found therein.

Any person who refuses admission on demand to a police officer or constable so authorized, or obstructs such officer or constable in the execution of his duty under this section, or who refuses on demand to disclose his name or address, or gives a false name or address, shall be liable to a penalty not exceeding five pounds.

Prosecution of offences and recovery of penalties in England.

14. In England, offences against this act may be prosecuted and penalties under this act recovered before a court of summary jurisdiction in manner directed by the Summary Jurisdiction Act.

In England "Summary Jurisdiction Act" means the act of the session of the eleventh and twelfth years of the reign of her present Majesty, chapter forty-three, intituled "An Act to facilitate the performance of the duties of justices of the peace out of sessions within England and Wales with respect to summary convictions and orders," and any act amending the same.

"Court of summary jurisdiction."

"Court of summary jurisdiction" means and includes any justice or justices of the peace, metropolitan police magistrate, stipendiary or other magistrate, or officer, by whatever name called, exercising jurisdiction in pursuance of the Summary Jurisdiction Act: Provided that the court when hearing and determining an information under this act shall be constituted either of two or more justices of the peace in petty sessions, sitting at a place appointed for holding petty sessions, or of some magistrate or officer sitting alone or with others at some court or other place appointed for the administration of justice, and for the time being empowered by law to do alone any act authorized to be done by more than one justice of the peace.

Power of offender in England to elect to be tried on indictment, and not by summary jurisdiction.

15. In England, where a person is accused before a court of summary jurisdiction of any offence against this act in respect of which a penalty of more than five pounds can be imposed, the accused may, on appearing before the court of summary jurisdiction, declare that he objects to being tried for such offence by a court of summary jurisdiction, and thereupon the court of summary jurisdiction may deal with the case in all respects as if the accused were charged with an indictable offence and not an offence punishable on summary conviction, and the offence may be prosecuted on indictment accordingly.

Form of appeal to quarter sessions.

16. In England, if any party thinks himself aggrieved by any conviction made by a court of summary jurisdiction on determining any information under this act, the party so aggrieved may appeal therefrom, subject to the conditions and regulations following:

(1.) The appeal shall be made to the next court of general or quarter sessions for the county or place in which the cause of appeal has arisen, holden not less than twenty-one days after the decision of the court from which the appeal is made; and

(2.) The appellant shall, within ten days after the cause of appeal has arisen, give notice to the other party and to the court of summary jurisdiction of his intention to appeal, and of the ground thereof; and

(3.) The appellant shall, within three days after such notice, enter into a recognizance before a justice of the peace, with two sufficient sureties, conditioned personally to try such appeal, and to abide the judgment of the court thereon, and to pay such costs as may be awarded by the court, or give such other security by deposit of money or otherwise as the justice may allow; and

(4.) Where the appellant is in custody the justice may, if he think fit, on the appellant entering into such recognizance or giving such other security as aforesaid, release him from custody: and

(5.) The court of appeal may adjourn the appeal, and upon the hearing thereof they may confirm, reverse, or modify the decision of the court of summary jurisdiction, or remit the matter to the court of summary jurisdiction with the opinion of the court of appeal thereon, or make such other order in the matter as the court thinks just, and if the matter be remitted to the court of summary jurisdiction the said last-mentioned court shall thereupon rehear and decide the information in accordance with the order of the said court of appeal. The court of appeal may also make such order as to costs to be paid by either party as the court thinks just.

Prosecution of offences and recovery of penalties in Scotland.

17. In Scotland, offences against this act may be prosecuted, and penalties under this act recovered, under the provisions of the Summary Procedure Act, 1864; or if a person accused of any offence against this act in respect of which a penalty of more than five pounds can be imposed, on appearing before a court of summary jurisdiction, declare that he objects to being tried for such offence in the court of summary jurisdiction, proceedings may be taken against him on indictment in the Court of Justiciary in Edinburgh or on circuit.

Every person found liable in any penalty or costs shall be liable in default of immediate payment to imprisonment for a term not exceeding three months, or until such penalty or costs are sooner paid.

Prosecution of offences and recovery of penalties in Ireland.

18. In Ireland, offences against this act may be prosecuted and penalties under this act recovered in a summary manner, subject and according to the provisions with respect to the prosecution of offences, the recovery of penalties, and to appeal of the Petty Sessions (Ireland) Act, 1851, and any act amending the same, and in Dublin of the acts regulating the powers of justices of the peace or of the police of Dublin metropolis. All penalties recovered under this act shall be applied in manner directed by the Fines (Ireland) Act, 1871, and any act amending the same.

Power of offender in Ireland to elect to be tried on indictment, and not by summary jurisdiction.

19. In Ireland, where a person is accused before a court of summary jurisdiction of any offence against this act in respect of which a penalty of more than five pounds can be imposed, the accused may, on appearing before the court of summary jurisdiction, declare that he objects to being tried for such offence by a court of summary jurisdiction, and thereupon the court of summary jurisdiction may deal with the case in all respects as if the accused were charged with an indictable offence and not an offence punishable on summary conviction, and the offence may be prosecuted on indictment accordingly.

Interpretation of "the Secretary of State" as to Ireland.

20. In the application of this act to Ireland the term "the Secretary of State" shall be construed to mean the Chief Secretary to the Lord Lieutenant of Ireland for the time being.

Prosecution only with leave of Secretary of State.

21. A prosecution under this act against a licensed person shall not be instituted except with the assent in writing of the Secretary of State.

Not to apply to invertebrate animals.

22. This act shall not apply to invertebrate animals.

INDEX.

LONDON: PRINTED BY C. F. HOWORTH, GREAT NEW STREET, FETTER LANE.